History, Imperialism, Critique

This book examines anti-imperialist thought in European philosophy. It features an international group of both emerging and established scholars who directly respond to Timothy Brennan's far-reaching call to rethink intellectual histories, literary histories, and the reading habits of postcolonialism, in relation to the anti-imperialist tradition of critique. Each contributor rethinks postcolonial and world literature, Continental thought, and intellectual history in relation to anti-imperialist histories and traditions of critique, through geographically diverse analysis.

This book provides a forum for the next generation of scholars to draw on and engage with the marginal yet influential work of the first generation of dissidents within postcolonial studies. It will appeal to researchers and students in the fields of postcolonial studies, world literature, geography, and Continental thought.

Asher Ghaffar is a PhD candidate in Social and Political Thought at York University, Canada. His research monograph, *Muslims in World Literature: Political Philosophy and Continental Thought*, is forthcoming with Routledge in 2019. His most recent essay on Zulfikar Ghose and Hanif Kureishi will appear in *The Routledge Companion to Pakistani Anglophone Writing*, edited by Aroosa Kanwal and Saiyma Aslam. He is also working on a second poetry collection.

Routledge Research in New Postcolonialisms
Series Editor: Mark Jackson, Senior Lecturer in Postcolonial
Geographies, School of Geographical Sciences, University of
Bristol, UK.

This series provides a forum for innovative, critical research into the changing
contexts, emerging potentials, and contemporary challenges ongoing within postcolonial studies. Postcolonial studies across the social sciences and humanities
are in a period of transition and innovation. From environmental and ecological
politics, to the development of new theoretical and methodological frameworks
in posthumanisms, ontology, and relational ethics, to decolonizing efforts against
expanding imperialisms, enclosures, and global violences against people and
place, postcolonial studies are never more relevant and, at the same time, challenged. This series draws into focus emerging transdisciplinary conversations that
engage key debates about how new postcolonial landscapes and new empirical
and conceptual terrains are changing the legacies, scope, and responsibilities of
decolonising critique.

Postcolonialism, Indigeneity and Struggles for Food Sovereignty
Alternative food networks in the subaltern world
Edited by Marisa Wilson

Coloniality, Ontology, and the Question of the Posthuman
Edited by Mark Jackson

Unsettling Eurocentrism in the Westernized University
Edited by Julie Cupples and Ramón Grosfoguel

History, Imperialism, Critique
New Essays in World Literature
Edited by Asher Ghaffar

For more information about this series, please visit: www.routledge.com/
Routledge-Research-in-New-Postcolonialisms/book-series/RRNP.

History, Imperialism, Critique

New Essays in World Literature

Edited by Asher Ghaffar

Routledge
Taylor & Francis Group

LONDON AND NEW YORK

First published 2019
by Routledge
2 Park Square, Milton Park, Abingdon, Oxon OX14 4RN

and by Routledge
52 Vanderbilt Avenue, New York, NY 10017

First issued in paperback 2020

Routledge is an imprint of the Taylor & Francis Group, an informa business

British Library Cataloguing-in-Publication Data
A catalogue record for this book is available from the British Library

Library of Congress Cataloging-in-Publication Data
A catalog record has been requested for this book

ISBN 13: 978-0-367-58732-1 (pbk)
ISBN 13: 978-1-138-21750-8 (hbk)

Typeset in Times New Roman
by Swales & Willis Ltd, Exeter, Devon, UK

Contents

Contributors

Himani Bannerji was born in Bangladesh in 1942 when it was still part of India. She was educated (BA, MA) and taught in Calcutta prior to immigrating to Canada in 1969. Her thesis (PhD, University of Toronto, 1988) is entitled: *The Politics of Representation: A Study of Class and Class Struggle in the Political Theatre of West Bengal*. Bannerji is an associate professor in the Department of Sociology at York University in Toronto. She is perhaps best known for her non-fiction writing in areas such as feminism, racism, and multiculturalism.

Eric Brandom is a James Carey Research Fellow in the History Department at Kansas State University.

Timothy Brennan is the Samuel Russell Chair in the Humanities at the University of Minnesota, and the author of foundational essays in postcolonial theory. His essays on intellectuals, imperial culture, and the politics of popular music have appeared in numerous publications, including *The Nation*, the *Times Literary Supplement*, *New Left Review*, *Critical Inquiry*, and the *London Review of Books*. He most recently authored the first of a two-volume study, *Borrowed Light: Vico, Hegel, and the Colonies*, with the second volume forthcoming. He is currently finishing an intellectual biography of Edward W. Said.

Sreya Chatterjee is Assistant Professor of English at the University of Houston. She specializes in twentieth-century British, Irish, South Asian, and postcolonial literature and feminist and materialist theories. Her current book project, tentatively titled *Narratives of Fracture*, studies core–periphery relationships in the postcolonial and globalization eras in twentieth-century domestic fiction. She has published essays and reviews in venues including *Comparative Literature Studies* and *Studies on Asia*. Most recently her essay on writer/activist Mahasweta Devi appeared in an edited volume *Naxalism: Poststructuralist, Postcolonial and Subaltern Perspectives* (Setu, 2017).

Daniel Dooghan is Associate Professor of World Literature at the University of Tampa, where he teaches courses on modern Chinese literature, postcolonial studies, and globalization. His recent publications include articles on

neoliberalism in video games and China's challenges to a singular world literature. He is currently completing a book manuscript on the literary configurations of a Chinese world-system.

Christian Gerzso is Visiting Assistant Professor of Interdisciplinary Studies at
Pacific Lutheran University. His work has appeared in *Textual Practice* and
The Battersea Review.

Asher Ghaffar is a PhD candidate in Social and Political Thought at York
University. His research monograph, *Muslims in World Literature: Political
Philosophy and Continental Thought*, is forthcoming with Routledge in 2019.
His most recent essay on Zulfikar Ghose and Hanif Kureishi will appear in *The
Routledge Anthology to Pakistani Anglophone Writing: Origins, Contestations,
New Horizons*, edited by Aroosa Kanwal and Saiyma Aslam. He is also work-
ing on a second poetry collection.

Mela Jones Heestand earned her PhD from the University of California, Davis in
Comparative Literature in 2014. Her area of specialization is South American
and French Caribbean literature. Currently, Mela lives in the Portland, Maine
area and works as a writer and independent scholar.

Auritro Majumder is Assistant Professor of English at the University of
Houston. His research and teaching interests include materialism and anticolonial thought.

Marco Katz Montiel has a PhD from the University of Alberta, Canada. He
publishes in English and Spanish on literature and cultural studies and is a
trombonist and composer.

Benjamin Noys is Professor of Critical Theory at the University of Chichester. He
is the author of *Georges Bataille: A Critical Introduction* (2000), *The Culture
of Death* (2005), *The Persistence of the Negative: A Critique of Contemporary
Theory* (2010), and *Malign Velocities: Accelerationism and Capitalism* (2014),
and editor of *Communization and Its Discontents* (2011).

Djordje Popović is a PhD candidate in Comparative Literature at the
University of Minnesota. In his dissertation, "The Concept of Statelessness
in Second World Literature," Popović examines the link between the
collapse of state socialism and the aesthetic of homelessness in Eastern
European literary modernism.

Introduction

Asher Ghaffar

The idea that world literature is constituted by a struggle between nations is often forgotten in those theories of postcolonial and world literature that arrive at an unfettered cosmopolitanism. No matter how much one tries to get beyond it, the nation haunts the historical imagination and disturbs ideas of world literature and global belonging. Despite world literature's renewed relationship with materialist and philological traditions, twentieth-century *worldly* anticolonialism is still a spectral presence within the field.[1] One of the unspoken reasons why twentieth-century anticolonialism arouses such uncertainty and distrust is because its corpus disturbs the neat oppositions between the nation and world, realism and modernism, religion and oral culture, etc. In fact, the conceptual contours of twentieth-century anticolonialism can be understood through such seismic junctures as the Russian Revolution (1917), the Mexican Revolution (1910), the Bandung Conference (1955), and the burgeoning anticolonial literary and cultural institutions that arose as a result. In spite of these, twentieth-century anticolonial struggles for sovereign nationhood and self-determination are overwhelmingly consigned to the nation-building past as part of the Eurocentric project in postcolonial and world literature.

These fields are defined by a break from Continental thought—not only a Marxism that is characterized as Eurocentric—but also a worldly humanism, which was developed most recently by the Palestinian-American intellectual, Edward Said. As an intellectual caught between generations, Said's remarkable achievement was to critique humanism from within when New Criticism and Textuality were the dominant modes of interpretation (Brennan *Edward Said*). Said's engagement with the Neapolitan, Giambattista Vico, was instrumental to his opposition to both camps during the Reagan era. Interpretation for Said was far from the autocratic act of the solitary individual, but a democratic attempt to understand the departures that the text often silences in its presumption of beginning. The present volume begins from another point than the field's early attempts to arrive at a "new" internationalism (Derrida) and offers the first long overdue engagement with the theoretical and literary contributions of Timothy Brennan in postcolonial studies and expanding our understanding of his insights in intellectual history and literary studies. The contributors begin the project of restoring the Vichian anticolonial tradition that

Brennan defends, while showing its relevance and critical edges in a number of literary and theoretical contexts.

Although Derrida attempts to reconcile with Marxism after the fall of the Berlin Wall by evoking an enigmatic internationalism "without status, without title, and without a name . . . without county, without national community" (Derrida 3–49), he does not provide a historical basis for his discussion. Homi Bhabha draws on his abstract internationalism when he writes: "What is striking about the 'new' internationalism is that the move from the specific to the general, from the material to the metaphoric, is not a smooth passage of transition and transcendence" (8). This "new" internationalism is opposed to an allusive old internationalism that is only mentioned in passing in relation to the grand narratives of class and capitalism that—as Bhabha claims—no longer structure the particular struggles of sexual and racial minorities.

While such postcolonialists consequently "updated" Saidian thought through colonial discourse analysis—emphasizing indeterminacy and paradoxically reinforcing identity—Marxists challenged his engagement with so-called conservative humanists such as Erich Auerbach and Leo Spitzer. Thus, Said was sometimes cast as the perennially conflicted "oriental" whose incompatible positions—and particularly whose *humanism*—were the anathema of postcolonialists and Marxists alike. However, what if, as Brennan suggests, Said's Vichian humanism was his major achievement? And what if it is precisely the relationship between humanism and Marxism that is *the* mark of "world" literature's twentieth-century's beginnings not only as a counter-canon to the oft-cited works in the field, but also as a philosophical outlook from the peripheries?

The contributors to *History, Imperialism, Critique: New Essays in World Literature* contend that it is precisely this relationship between humanism and Marxism that brings twentieth-century anticolonial works into sharper focus both as a philosophy and as a literary form. In making this connection, the contributors draw on Brennan's insights in postcolonial studies and develop them into a number of understudied literary, cultural, and national contexts. As opposed to seeing twentieth-century anticolonialism as rupturing Continental thought, Brennan has consistently traced these philosophical affiliations and nuanced the distinctions in powerful and far-reaching critiques. In *Borrowed Light: Vico, Hegel, and the Colonies* (*Borrowed Light*) he puts forward an alternative to the false logic that sets Vichian humanism against Marxism in an utterly compelling theory of twentieth-century interwar anticolonialism. In *Borrowed Light*, Brennan revises the view of Said who saw the Neapolitan as distinct from Marx. In a ground-breaking reading of Vico, he argues that anticolonialists reinterpreted the Western Marxist tradition and Continental thought in a particular way that built on this Vichian philosophical linkage, which provides the unacknowledged nexus point between Western Marxism and anticolonialism. This is particularly evident in Vico's prescient proto-Marxist social theory that anticipates the anti-imperialist connection between class conflict and foreign domination. The philologist's outlook also resonates with anticolonial thought in his dethroning of the Greeks, his emphasis on orature and myth as the starting points for the study of human history, and his foregrounding of mimetic intelligence as a challenge to Descartes' rationalism. The counter-cast of figures is equally important to Brennan, who provides critical readings of the colonialism implicit in Nietzsche, Spinoza, and others who provide the contrast to Vico. This is not to say twentieth-century anticolonialists all read Vico, or simplistically opposed their ideas with those of Nietzsche; rather, the work of these thinkers illuminates the conceptual coordinates for studying the twentieth-century interwar anticolonial corpus *as* world literature.

The approach to the study of anticolonialism in this anthology stands in contrast to the contemporary emphasis on identity and difference that coincides with the erosion of the possibility of the social. In this enclosing space, the religious subject returns with a vengeance as the reified agent of a social change. Moreover, the elusive intellectual histories that allow for a rethinking of anticolonialism—which such fundamentalisms have eroded—are not frequently discussed in critical theory, postcolonial studies, or world literature. Vico's foregrounding of myth and orature stretches dialectical thought into peripheral contexts—where religion and oral culture dominate as social forms. Adorno, by contrast, saw any conscious attempt to represent religion as a superimposition on the artwork and not part of the actually existing social world, which had already overcome religion through instrumental rationality. Despite his insight, religion still persists in the contemporary world and has vehemently returned. In peripheral and semi-peripheral contexts, religion in many ways is still central to social relations and thus must be taken seriously to rethink peripheral aesthetics.

In a critical sense, religion and oral culture contain within them a demand for another world that, when placed in tension to the actually existing world, does not only stand helpless against it; if it does, it betrays that hope, or falls back on false universals such as civilization, ethnicity, tribe, or class interest. In this demand is a promise that is neither of this world, nor beyond it. This idea is in keeping with a Vichian outlook to rethink myth in world literature, which, as Wilson Harris suggests in the West Indian context, generates "nebulous" links in "a latent ground of the old and new" (qtd. in Bundy 140). Without considering the relationship between oral culture and religion, it is virtually impossible to rethink anticolonialism and its intimate relationship to Marxism in peripheral states. As the contributors to this anthology demonstrate, these residual and active elements need to be put into relationship with modernity to develop a viable peripheral theory of literature.

Brennan's work takes twentieth-century anticolonialism seriously as an intellectual tradition in its own right and does not liquidate Continental thought in search of radical cultural difference as is the case in Subaltern Studies and decolonial thought (see, for example, Chatterjee and Mignolo). Thus what is at stake in his reading is the meaning of anticolonialism that he conceives philosophically—linking it to the Hegelian Marxist tradition through a novel reading of Vico. His reading would add tremendously to the archive of thinkers, literary works, and languages that are usually drawn on in world literature. Brennan contends that the interwar period (1918–1939) was a crucible for the contradictions that continue to shape cultural and literary theory today. However, despite the profound impact of Left Hegelianism on anticolonial thought, much postwar theory excises the interwar moment with major ramifications for intellectual history, literary history, and the reading habits that inform the field (Brennan *Marxism, Modernity*). Brennan's view of literary borrowing challenges the insidiously persistent notion in postcolonial and world literature that so-called nationalist works are derivative of Western modernist classics, content-driven, or simply Eurocentric—that is, not *all* borrowed forms, which require philological reconstruction—even in their unevenness and deceptive simplicity.

As a contrast to this view, Pascale Casanova's seminal Bourdieu-inflected theory examines how world literature enters a global literary market through modernist "literary capital." However, she overlooks the ways that social realism was international and also drew on such avant-garde forms as montage (12–21). When viewed from a world scale, Brennan brings together seemingly disparate aesthetic categories, such as montage and polemic, to demonstrate how such categories were mediated throughout the interwar period in peripheral aesthetics. Moreover, the far-reaching effects of the Communist International on the anticolonial internationalism are lost in Casanova's discussion. The great black American novelist, Richard Wright, provides one of many illuminating examples of a social realist writer who gained international recognition due to his association with the communist party. Wright does not explicitly draw from the modernist forms in his internationally recognized novels—as he writes:

> Out of step with our times, it was but natural for us [writers] to respond to the Communist party, which said: "Your rebellion is right. Come with us and we will support your vision with militant action." Indeed, we felt that we were lucky. Why cower in towers of ivory and squeeze out private words when we had only to speak and millions listened? Our writing was translated into French, German, Russian, Chinese, Spanish, Japanese . . . Who had ever, in all human history, offered to young writers an audience so vast? True, our royalties were small or less than small, but that did not matter.
>
> (Wright in Denning 63)

In *The Blueprint for Negro Writing*, Wright suggests that black writers must accept nationalism in order to go beyond it. Furthermore, the poet Amiri Baraka did not receive international recognition in the American modernist tradition; but his symbolic move to Harlem, after the death of Malcolm X, marked the beginning of the Black Arts Movement—a nationalist and international literary movement whose print culture arose in Harlem and not Paris.[2]

In light of political expediency, the cultural critic Michael Denning claims that aesthetic form was never a priority for anticolonial writers. Denning's distant readings provide an exciting, panoptic, but ultimately conceptually unclear sense of the commonalities across twentieth-century anticolonialism. *Borrowed Light*, by contrast, puts forward concepts from an anticolonial terrain to reconstruct the feel of the time—as the historical excess is lost in the onward march of postwar theory into increasingly specialized fields such as post-humanism. Indeed, as this volume demonstrates, the conceptual dimensions of monumental works by poets as diverse as Aimé Césaire and Rabindranath Tagore can now be understood in unexpected philosophical relationship. Many other works, such as the stream-of-consciousness Urdu novel *London ki Ek Raat* (*A Night in London*) by the socialist writer Sajjād Ẓahīr, have an incomplete quality, which present the reader with an interpretive challenge. Such works demonstrate that the oppositions of realism and modernism, for instance, were not as facile as they later became in postwar literary history. That is, what is interesting is the incomplete quality of these works that speak to their lack of integration of modernist aesthetics such as stream-of-consciousness.

Brennan asks us to consider this incomplete quality and the borrowing of diverse forms as a challenge, and his return to close reading aligns him more with the overlooked Brazilian literary critic, Roberto Schwarz. In his unprecedented *A Master on the Periphery of Capitalism: Machado De Assis*, the literary theorist redefines immanent critique by examining how European modernist ideas appear dissonant in a Brazilian social structure whose social and economic base was slavery. Schwarz's readings build on "the contradictory traditions of the Frankfurt School and 'the inspiration of Marx'" (4). Adorno and Lukács are particularly important in Schwarz's reading of formal and ideological dissonances: "Immanent criticism of intellectual and artistic phenomena pursues the logic of its aporias . . . in such antimonies criticism perceives those of society" (Adorno, *Prisms* 32). When borrowing is resituated in another socioeconomic context, its antimonies are exposed, as are the broader contradictions that exist between the nation and world. Consequently, Schwarz's reading also challenges the problems plaguing American literary criticism and its opposition between the nation and the world.

Brennan's rethinking of anticolonial intellectual history cannot be characterized as a framework. It is rather a rethinking of the dialectical tradition that is not a procedure, a method, or something that can be applied a priori to the literary and cultural work, but understood in relation to the "felt contact with . . . objects" (Adorno, *Reflections* 247) which internalize the antimonies of the social world. As such, through close readings, the contributors grasp how anticolonial thought developed—not as it is retrospectively reinterpreted by postwar theory. What makes Brennan's rereading of the anticolonial tradition intriguing from the perspective of Continental thought is echoed by Adorno: "At the present time Hegelian philosophy, and all dialectical thought, is subject to the paradox that it has been rendered obsolete by science and scholarship while being at the same time more timely than ever in opposition to them" (*Hegel* 55).

Brennan's far-reaching and revolutionary insight in *Borrowed Light* is in situating Vico within the anticolonial tradition. To take this further, if the rediscovery of Asia through a textual revolution resulted in a second renaissance of humanism in Europe (Schwab), the reception of Continental thought and the *reinterpretation* of the human through twentieth-century anticolonialism marks a *third moment* in a humanist renaissance from the peripheries.

What this means is virtually indisputable as far as the historical record goes. Many major anticolonialists sought to stretch dialectical thinking by redefining the concept of the human from within—often in relationship to a counter-cast of philosophers such as Nietzsche and Heidegger who overshadow the "beginnings" of postwar theory.

It is time to return to twentieth-century anticolonialism as we find ourselves within something like this moment. The sociologist Albert Memmi, who once spoke for anticolonialism, now views it belligerently and compares postcolonial nations to sick children whose benevolent parents, imperial states, watch over them: "The decolonized nations are like the children of aging parents, born weak and suffering, the fruit having dried before it had the chance to mature" (Memmi 54). Notwithstanding Frantz Fanon's critique of the colonized middle-class, there are obvious reasons to emphasize the failures of twentieth-century anticolonialisms. However, overlooking interwar anticolonialism means that the histories of struggle that continue to shape the world today are lost at the level of theory—for anticolonialism is also a theory that overwhelmingly draws from the Vichian and Marxist categories that Brennan lays out in *Borrowed Light*.

The contributors of the present volume continue Brennan's necessarily collective project of reconstructing an anticolonial international republic of letters and rightly see the world within the nation and the nation within the world. *Borrowed Light* reconceptualizes anti-imperialist literature and traditions of critique in relation to Continental thought to challenge and reconstruct intellectual history, widen the scope of literary history, and put forward another vocabulary for studying world literature. Accordingly, the first section of *History, Imperialism, Critique* provides critical reappraisals of anti-imperialist intellectual history through such figures as Aimé Césaire and Rabindranath Tagore; the second section delves into literary history, examining literary histories not normally included in the postcolonial canon, such as that of China; and the third section, poetic history, focuses on literary works, providing close readings of those not included in the postcolonial and world literature canons. Deriving their readings from literary and cultural works rather than the inherited reading habits that continue to dominate the field, the contributors implicitly make the case for an anticolonial world literature.

The chapters

Auritro Majumder situates the anthology in the field of world literature and considers the forgotten intellectual and philosophical works of Mao Zedong and Rabindranath Tagore who he argues helped to consolidate the concept of the 'East' and emphasized culture as the site of anticolonial resistance. The Marxist, antiracist, and feminist sociologist Himani Bannerji examines the implicit materialism in Tagore's modernism and argues that it is through his humanism that decolonization becomes possible. In Chapter 4, Asher Ghaffar reconsiders the overlooked impact of the Western philosophical dimensions of Muhammad Iqbal's social and political thought to demonstrate how his engagement with British Hegelianism and his immanent critique of Nietzsche can be best understood in relationship to the Second

Boer War and the Russian Revolution respectively. Eric Brandom in Chapter 2 locates in *Cahier d'un retour au pays natal* and Césaire's other work an "insurgent and disruptive humanism" (30) that bears striking affiliations with Vico's ideas. As a whole, this section demonstrates the importance of Vico's humanism for reconstructing a worldly anticolonial intellectual history from the Edwardian to the interwar period.

The next section of the anthology turns to literary history and three overlooked contexts: China, the Soviet Union, and Mexico. In Chapter 5, Daniel Dooghan challenges the interpretive traditions that view Lu Xun as either a national allegorist or modernist. He argues that both readings suppress the local dimensions of his thought, particularly the vernacular, which can be recuperated through a philological reading of Lu Xun's works. Djordje Popović in Chapter 6 considers how the major communist author, Platonov, was refashioned into a social realist and modernist. He unpacks the fallacies of such a dichotomy by exposing the Heideggerian assumptions about language that underpin Brodsky's modernist reading of Platonov, which have come to shape his reception in world literature. In Chapter 7, Christian Gerzso distinguishes the post-revolutionary Mexican avant-garde *Estridentismo* movement from its European counterparts by drawing on Brennan's anticolonial categories such as montage, unevenness, sacrifice, and polemic. In addition, he suggests that the avant-garde group's irreverent deployment of irony allows us to expand Brennan's conception of peripheral aesthetics in this context. Among other things, this section demonstrates how the modernist and social realist dichotomy, which also shaped the development of Marxist interwar aesthetics, eventually solidified into potent interpretive myths that recirculate in the field of world literature.

In the final section, "poetic history," the contributors show that theories need to arise from close readings of individual works rather than the other way around. Sreya Chatterjee in Chapter 8 examines the Irish context. Through a novel reading of Brian Friel's play *Translations*, she demonstrates how Friel's rendition of colonialism in the Irish rural context in the 1830s emphasizes the popular dimension of culture, which existed in opposition to a hegemonic nation. She contends that the play reflects on this popular cultural configuration, which defined itself through a Vichian refashioning of the vernacular. Similarly, in Chapter 9 Montiel considers the influence of Vico and Hegel on reading Rodó's *Ariel*, particularly by first highlighting Washington's expanding empire, and Rodó's response, which drew on Vico, Hegel's concept of spirit as a manifestation of the will of the people, and a cosmopolitanism to challenge the pitfalls of nationalism. In Chapter 10 Mela Jones Heestand draws our attention to José Arguedas' novel, *Deep Rivers* (1957), which was considered to be a temporally regressive regional novel by those who recuperated it during the Latin American "boom" era. Heestand instead emphasizes Arguedas' (Vichian) non-presentist concept of time, which prompts her to unpack the novel's Quechua worldview and align it with Voloshinov's future-oriented view of language. In Chapter 11, the critical theorist Benjamin Noys considers the global reach of Nietzsche and poses the important question of how to mediate his reception in postcolonial studies as either an anticolonialist or a colonialist. Through close readings of novels by Roberto Arlt, Waguih Ghali, and Sam Selvon, Noys suggests that these novelists "impersonate" Nietzsche, drawing on such concepts as the over-man; in doing so, the postcolonial novel becomes the site to push critique beyond the aforementioned receptions of Nietzsche. Timothy Brennan completes the anthology by and reclaiming the historical and emancipatory possibilities of Vico's idea of poetic history in relation to other twentieth-century iterations of poetics.

Notes

1 Many have argued that the field of postcolonial studies sought to displace anticolonialism—particularly when the latter was Marxist in orientation (see Lazarus; Larsen; Parry). The popular post-Orientalist writer, William Dalrymple, writes, for example, that postcolonialists have rarely examined the archives that document the Indian Rebellion of 1857: "For a time when ten thousand dissertations and whole shelves of Subaltern Studies have carefully and ingeniously theorized about orientalism and colonialism and the imagining of the other . . . not one major study has ever been written from the Mutiny Papers, no major study has ever systematically explored its contents" (Dalrymple 13–14).
2 For more on the print culture in twentieth-century Harlem, see Brent Hayes Edwards, *The Practice of Diaspora: Literature, Translation, and the Rise of Black Internationalism*. Cambridge, MA: Harvard UP, 2003.

Works cited

Adorno, Theodor. *Hegel: Three Studies*. Translated by Shierry Weber Nicholsen. MIT Press, 1994.
—. *Minima Moralia: Reflections on a Damaged Life*. Verso, 2005.
—. *Prisms*. Translated by Shierry Weber Nicholsen and Samuel Weber. MIT Press, 1981.
—. "Theses upon Art and Religion Today." *The Kenyon Review*, vol. 18, no. 3/4, 1996, pp. 236–240.
Ahmad, Aijaz. *In Theory: Nations, Classes, Literatures*. London: Verso, 2008.
Auerbach, Erich. *Mimesis: The Representation of Reality in Western Literature*. Princeton UP, 2003.
Bartolovich, Crystal, and Neil Lazarus, eds. *Marxism, Modernity and Postcolonial Studies*. Cambridge UP, 2002.
Bhabha, Homi. *The Location of Culture*. Routledge, 2004.
Brennan, Timothy. *Borrowed Light: Vico, Hegel, and the Colonies*. Stanford UP, 2014.
—. "Edward Said and Comparative Literature." *Journal of Palestine Studies*, vol. 33, no. 3, 2004, pp. 23–37.
—. "Postcolonial Studies between the European Wars: An Intellectual History." *Marxism, Modernity and Postcolonial Studies*. Edited by Crystal Bartolovich and Neil Lazarus, Cambridge UP, 2002, pp. 185–203.
Bundy, Andrew, ed. *Selected Essays of Wilson Harris*. Routledge, 2005.
Casanova, Pascale. *The World Republic of Letters*. Harvard UP, 2007.
Chatterjee, Partha. *The Nation and Its Fragments: Colonial and Postcolonial Histories*. Princeton UP, 1993.
Dalrymple, William. *The Last Mughal*. Vintage, 2007.
Denning, Michael. *Culture in the Age of Three Worlds*. Verso, 2004.

Derrida, Jacques. *Specters of Marx: The State of the Debt, the Work of Mourning, and the New International.* Routledge, 1994.

Dirlik, Arif. "Where Do We Go From Here?" *Marxism, Modernity, and Postcolonial Studies Diaspora,* vol. 12, no. 3, 2003, pp. 2–19.

Fanon, Frantz. *The Wretched of the Earth.* Trans. Richard Philcox. Grove Press, 2004.

Larsen, Neil. *Determinations: Essays on Theory, Narrative, and Nation in the Americas.* Verso, 2001.

Lazarus, Neil. *Nationalism and Cultural Practice in the Postcolonial World.* Cambridge UP, 1999.

Lukács, Georg. *The Historical Novel.* Merlin Press, 1962.

Memmi, Albert. *Decolonization and the Decolonized.* U of Minnesota Press, 2006.

Mignolo, Walter. *The Darker Side of Western Modernity: Global Futures, Decolonial Options.* Duke UP, 2011.

Parry, Benita. *Postcolonial Studies: A Materialist Critique.* Routledge, 2004.

Said, Edward. *Beginnings: Intention and Method.* Basic Books, 1975.

Schwab, Raymond, Patterson-Black, Gene, and Reinking, Victor. *The Oriental Renaissance, Europe's Rediscovery of India and the East 1680–1880.* Translated by Gene Patterson-Black and Victor Reinking. Columbia UP, 1984.

Schwarz, Roberto. *A Master on the Periphery of Capitalism: Machado De Assis.* Ed. John Gledson. Duke UP, 2001.

Vico, Giambattista. *New Science.* Penguin Books, 1999.

Ẓahīr, Sajjād. *A Night in London.* Edited and translated by Bilal Hashmi. Harper Perennial, 2011.

Part I
Intellectual history

1 World literature, the *Geist*, and the East, 1907–1942

Auritro Majumder

> The utopian impulse in thinking is all the stronger, the less it objectifies itself as utopia . . . Whatever was once thought . . . can be suppressed; it can be forgotten and can even vanish. But it cannot be denied that something of it survives.
>
> (Theodor Adorno, 'Resignation' 168)

I

While the idea of a transnational republic of letters has a long European history, memorably expressed by Dante in the fourteenth, Erasmus and More in the sixteenth, and Goethe in the nineteenth centuries, it is only in the twentieth century that its promise is realized and, more crucially, *displaced* on a world scale (Said). The latter includes such events as the establishment of the World Literature Publishing House in the Soviet Union in 1918, the All-India Progressive Writers Association in British India in 1936, and the Afro-Asian Writers Association founded in Sri Lanka in 1958 and finalized in Egypt in 1962. These institutions and the ideas they traced, albeit little recognized today, constitute a vast corpus of anticolonial aesthetic and literary theory. Disparate texts present aspects of this demotic humanist vision: such as the Indian Rabindranath Tagore's lecture on *visva-sahitya* (world literature) in 1907; the Hungarian Georg Lukács' 'Narrate or Describe' in 1936; the Chinese Mao Zedong's Yenan Forum talks on literature in 1942; the German Erich Auerbach's 'Philology and *Weltliteratur*' in 1952; and lastly, the Ngũgĩ wa Thiong'o-led Kenyan group's (with Taban Lo Liyong and Henry Owuor-Anyumba) 'On the Abolition of the English department' in 1972. I mention the nationalities of these writers for no other reason than to illustrate the transnational reach of their argument. Here, broadly speaking, one finds an emphasis on the coevality of imperial and colonized cultures; the focus on the oral, the vernacular, and the folk; the unity of thought and action; and lastly, the centrality of creative, purposive human agency in the task of social transformation. Extending the intervention of Brennan's *Borrowed Light* into recent debates in world literature, this essay engages with Tagore and Mao. I elucidate Rabindranath Tagore and Mao Zedong's innovative, if largely forgotten, contributions to the development of a twentieth-century Asian trajectory of dialectical criticism and philosophy of literature, and how these might nuance currently dominant configurations.

Academic literary criticism, exemplified by books such as Pascale Casanova's *The World Republic of Letters* and essays such as Franco Moretti's 'Conjectures on World Literature', has injected fresh vitality into the life of *weltliteratur*: a concept developed by Johann Wolfgang von Goethe in a number of essays and letters in the early nineteenth century. What is common to Casanova and Moretti, despite their different foci, is a shared interest in locating diverse national literatures within a world system: tracing transmission, production, circulation, and reception of literary forms and genres. To the extent that such discussions are rooted in a materialist understanding of style (and are not simply reiterations of the traditional model, describing the diffusion of stylistic influence from the metropole to the periphery), they represent a break with the past. By materialism, I mean that world literature scholarship by and large has moved beyond the 1980s and 1990s trends of literary and cultural theory: on one hand, the poststructuralist insistence on seeing literature as a randomly assembled order of signs, signifiers without signifieds, as well as the influential corollary; on the other hand, of postcolonial criticism that claimed that 'Western' conceptual tools were simply inadequate to deciphering the non-Western world, and as such any unitary (Enlightenment, humanist) model of interpretation perforce would have to be discarded in favour of a deconstructive, colonial discourse analysis. As I have argued elsewhere, nestled within the so-called 'transnational turn' in literary studies, what has returned, unanticipated and mostly unacknowledged, is *cultural materialism* (Majumder). In today's mix of distant and surface reading, world systems theory, and major and minor literary traditions, one traces "the experienced tensions, shifts and uncertainties, the intricate forms of unevenness and confusion" that mark "lived experience" and its mimetic representation in literature (Williams 131, 129).

To invoke Raymond Williams is to suggest that current theorizations of world literature have their roots in traditions of cultural materialism, best encapsulated in Williams's work and others such as the Brazilian Marxist critic Roberto Schwarz. It is also to say that such influences have been largely ignored, effaced, or misrepresented. Thus, Schwarz, undoubtedly one of the key theorists of dependency in Brazilian literary criticism, finds no mention in Pascale Casanova's recent discussion of dependency in the Latin American novel, nor in her much-quoted notion of the "Greenwich meridian of literary taste" (92–101, 232–234). In turn, Moretti, discussing the export of metropolitan literary forms of the novel and their interaction with diverse local contexts in the colonial and non-colonial peripheries, has this to say about Schwarz:

> Now, that the same configuration [of export of ideas] should occur in such different cultures as India and Japan—this was curious; and it became even more curious when I realized that Roberto Schwarz had independently discovered very much the same pattern in Brazil.

(58)

Schwarz's originality however is neither "independent," as Moretti wants us to believe, nor can Schwarz be adduced so easily to Moretti's (supposedly un-American) "distant reading" model as he claims (57–58). Roberto Schwarz, as anyone who has read him would know, belongs to the tradition of dialectical materialist criticism; one that includes his teacher at the University of Sao Paolo, the formidable literary critic Antonio Candido, as much as the key figures of interwar Western Marxism, Georg Lukács and Theodor Adorno (López). While Moretti sees Schwarz's dialectical method as the precursor to his own model of distant reading, I believe—and genuinely want to emphasize—that they are quite opposed, and further that such appropriations are less innocent and far more complex. The way current scholars read past critics has a lot to do with how genealogies of thought are legitimized: in other words, what is permissible, and what is censored as non-transferable and contrarian to the present order of knowledge. All of this is of course shaped by dominant ideologies as much as by the conditions of unevenness in global knowledge production. In that respect alone, one needs to be on guard against appropriation under the sign of interpretation: Adorno's pithy remark in *Minima Moralia*, "Advice to intellectuals: let no one represent you" (128), is rather relevant in this context.

To take Schwarz as an example: in his influential essay, 'Brazilian Culture', commenting on the problem of cultural 'imitativeness' in Brazil, Schwarz observes that "the painfulness of an imitative civilisation is produced not by imitation—which is present at any event—but by the social structure of the country" (15). Rather than seeing literary production in the nominally independent yet structurally dependent periphery (Brazil) as testifying to an ontological lack, a "copy" of an "original" as compared to European "norms," Schwarz sees Brazilian literature as articulating a *social* structure of unevenness, of *mixed* norms. Significantly, Raymond Williams theorized a similar "structure of feeling" in describing the changing literary role of the countryside in Britain. In Schwarz's study of nineteenth-century Brazil, a tiny Europeanized intellectual class imbued with modern ideas of liberty and progress struggled to express the social reality of a country undergirded by slave labour and an export-oriented plantation economy. The distinctive form of social relations in Brazil—constituted by an enlightened elite, an intermediate group of *agregados* or free men and women dependent on favours, and the directly enslaved—is part of capital's unevenly universalizing narrative. The condition of "backwardness" and "national dependency" in such a peripheral context is "of the same order of things as the progress of the advanced countries" in the metropolitan core (14); "our 'backwardness'," Schwarz claims, is "part of the contemporary history of capital and *its advances*" (16, original emphasis). Drawing, above all, from Adorno, Schwarz contends that it is literary form more than thematic content that best captures the contradictions of social and historical totality. As Schwarz shows in his study of Machado de Assis, the pre-eminent nineteenth-century novelist, the narrative structures of Machado's novels come to formally articulate the peculiar shape of Brazilian modernity. Narrativizing a society different from but also beholden to contemporary France or England in key aspects, Machado departs from the nineteenth-century conventions of European naturalism, as his work combines and takes on the dual character of modernist experiment with narrative form, and the realist critique of the bourgeois order.

Such departures need not be restricted to Brazil alone. The differential calibration of realism and modernism and aesthetics and politics in non-European, peripheral literatures will be familiar to current scholars even if many of them, curiously, are unfamiliar with Schwarz's work. What is less familiar today is Schwarz's *partisan* approach, in the old political sense of the word. For Schwarz suggests, rather emphatically, that the crisis of representation in the periphery can be resolved only through greater popular "access" to literature and "culture." The peculiar contradiction of peripheral representation rests "on the exclusion of the poor from the universe of contemporary culture" (16). This is a generalizable point about capitalist culture *tout court*, and not simply in the peripheries of the world system. It goes further than arguments about economic justice for the "poor": the dialectical suggestion is made that culture itself is also impoverished by being restricted to the "elite" and the "dominant class," who are necessarily locked into a condition of dependence on the metropolis since they are "the beneficiaries of . . . [the] given situation" (15). The poverty of culture as it exists goes beyond "cultural transplantation," Eurocentrism, or even ways of reading.

What Schwarz testifies to by way of dialectical criticism is the partisan nature of cultural activity, that which is immanent within the present but which is activated only by purposive labour, with the "workers gaining access" to their conditions of existence and "re-defin[ing] them through their own initiative" (16). Situated squarely within the realm of human practice, the outcomes of such processes are both local and global, "either promis[ing] . . . or catastrophic." The form of Schwarz's argument is itself dialectical: successively invoking nativist or indigenist, postmodern, and ontological reading practices (all dominant theoretical positions in the US academy, I should point out), only to reveal their inadequacy. If dialectical criticism is posited as a "contrar[ian]" method it emerges only in opposition to prior available options. Dialectics must fold within itself the present order of things—what Hegel describes in the *Phenomenology* as the schema of negation, fulfilment, and sublation—for another, *better* future to emerge; so must the literary critic.

What is left out of the current genealogy of world literature, arguably, is the concern with substantive rather than nominal democracy. Such a concern animates the thought of Goethe, writing in the shadow of the French Revolution, and equally occupies the thought of Adorno, Williams, and Schwarz—as well as Tagore and Mao who I discuss below—in the twentieth century. This is of course a non-representative list of thinkers (along gender lines, for example) in the cultural materialist tradition. A preoccupation with emancipation is a defining feature of these progressive intellectuals, and both these terms—*emancipation* and *progress*—have been complicated endlessly ever since, to the point where they have disappeared as respectable intellectual goals. If it is at all worthwhile to recount their intervention, especially, in registering world literature to the collectivist desire for emancipation, it is imperative to ask—at this point in the twenty-first century—what is *still relevant* in that move.

This is the crucial significance of Timothy Brennan's *Borrowed Light: Vico, Hegel, and the Colonies*, to which the present essay (and this volume) serves as a response. *Borrowed Light* argues that Vico nuances and in many ways prefigures Hegel's concept of the *Geist* or Spirit as the self-realization and actualization of human freedom. Brennan posits Vico as the largely unacknowledged master, politically and philosophically, of Western Marxism. In doing so Brennan, a student of Edward Said, provides a remarkable remapping of the roots of contemporary Marxist humanism and anticolonial thought. He draws our attention to a less-discussed aspect of Said's oeuvre, namely his affinity to Vico, especially the latter's insistence that the earliest human knowledge is expressed through oral myth and poetry (literature), and that it is the human imagination, at work, that grounds social and political change, rather than positivistic science or instrumental rationality. In foregrounding the political implications of debates in modern Western philosophy, *Borrowed Light* resituates one of Said's—and postcolonialism's—singular interventions, namely the role of literary and cultural representation in shaping, and not simply reflecting, imperial domination and resistance.

What we have taken to be a key methodological and conceptual point in Brennan is his reading of modernity, in terms of philosophy, as centred on the concept of *agon* or processive struggle. This provides a counterpoint to the more familiar Jamesonian method of assimilation, where distinct and discrete schools of poststructuralist theory, "anti-philosophy," and even Hegelian dialectics are sought to be reconciled in their difference (see Brennan 109–118). Drawing on a series of encounters between Spinoza/Vico and Hegel/Nietzsche, Brennan rightly argues that what is at stake, finally, in these admittedly abstruse and "Western" philosophical disputations is the most important political issue of the modern era—what it means to be *human*. Immanent in these philosophers' ways of viewing the world, and something of which they themselves were often acutely aware, is a methodological "war of positions": not a priori assumptions imposed from without, but a logical working out of inherent tensions and contradictions, and a processive evaluation of intellectual legacies and disavowals. As a *reading* practice, this is the core of Hegel's famed truth procedure: a challenge to read "negatively" without "presuppositions." Equally, if the Vichian-Hegelian lineage of the Enlightenment challenged Cartesian rationality and the Nietzschean return to revanchist notions of ontology—three distinct trajectories, very far from our settled opinions of a monolithic Enlightenment—it did so, as Brennan reminds us, through the category of "labour." This is labour understood, expansively, in the Hegelian sense of the term: as intentional, purposive human activity.

Such a Vichian-Hegelian conception is by no means limited to pre-twentieth-century Europe; rather, it impinged on the thought of anticolonial intellectuals such as Tagore and Mao in the first part of the twentieth century, and found its way into postcolonial theory through Said in the second half. As Tagore would say, this is the work of *sahitya* or literature. The rational thought activity characteristic of the *Geist* is neither in counterpoint to the imagination, nor is it reducible to the economic sphere in the figure of *homo economicus*. Focusing through a Vichian lens allows us to see the conceptual networks and lineages between

Western Marxist thought, with humanism and the collective oral imagination as its twin leitmotifs in Brennan's remarkable account, and anticolonial intellectuals and their equally Vichian critique of class and foreign domination. The theoretical frame of human thought-as-action continues to provide a figuration of political and aesthetic resistance in our supposedly posthuman, globalized times.

Borrowed Light's critique of the academic tendency to emphasize "rupture" over "continuity" (11–14) is applicable to at least two dominant strands of world literature scholarship. Current world literature scholarship either erases or misappropriates prior cultural materialist work. The first school, represented by critics such as Casanova and Moretti, accepts the *continuing* structuring role of imperial and capitalist unevenness as determinate conditions for literary production. Yet in the last instance, these critics' aim becomes descriptive—that is, an attempt to classify the social form of capitalist modernity in quasi-scientific, even pragmatic terms. This already represents a retreat from older, and robustly anti-positivist, traditions of materialism per se. The second tendency, exemplified by Susan Stanford Friedman's recent magisterial study, *Planetary Modernisms*, is even more ambitiously revisionist: it discards any notion of modernity as the historically specific and concrete social form of capital emerging on a world scale in the sixteenth century. Friedman, following Fernand Braudel's notion of *longue durée* and Bruno Latour's network theory, and drawing (incorrectly) from Gayatri Spivak's notion of planetarity, sees modernity and modernism everywhere and at every time, thus—in effect—nowhere. In such accounts of planetary modernity, the constitutive role of capital, especially colonial-capitalism, is relegated to a second order of explanation if not *discarded* wholesale. By posing the contemporary moment in terms of rupture and radical newness, the long struggle against capitalism, colonialism, patriarchy, and combinations thereof is marginalized in favour of the innocent exploration of new formalisms, digital media, ontology of things, and so on.

To emphasize by contrast a different, and better, kind of continuity is to go back to Adorno's observation about the "utopian impulse in thinking," quoted at the beginning of this essay. In his very brief piece 'Resignation', from which these remarks are taken, Adorno argued that the role of the intellectual is not to submit their critical autonomy to the collective cause no matter how convincing or appealing it may be (his own reference is to the events of 1968 and his non-participation in them). To reduce thought to an objective, that is, positive utopia is to resign one's critical faculty. While I have no desire to defend Adorno's attitude towards activism, the importance of his remarks lies in another direction. Adorno reminds us that thinking—thought-as-activity—is at its most potent and utopian when it continues to resist the historical and real-world defeats of its ideas. Resistance lies in the negation, in and through thought, of failed utopias as well as victorious reaction, whether these are of the anticolonial nationalist, or socialist, variety. It is necessary to repurpose his insight for the twenty-first century, especially for the loss of large-scale macro ideas that seems to plague academia. While the previous century's versions of anticolonialism and socialism have receded in our own time, their critique of the world as shaped by capital still remains, arguably, in the most part, valid. Rather than accepting the demise of emancipatory utopias, the proper task of present-day critique is to remain true to its negative charge. To explore this point further, I turn to the lectures by Rabindranath Tagore and Mao Zedong, and the insights of dialectical literary theory in the first half of the twentieth century.

II

1907 marks the year of the Indian poet and philosopher Rabindranath Tagore's lecture on *visva-sahitya* or 'world literature'. It is a significant date in literary history: Tagore spoke to the National Council of Education in Calcutta, a fledgling nationalist body that had been set up a year before to develop university education away from colonial control. *Visva-sahitya* or world literature was Tagore's term for 'comparative' literature, and arguably the first articulation of its kind in India and in the non-European world.

It is necessary to say a few words about the juncture, particularly in the Asian continent, in which Tagore's *visva-sahitya* takes shape. Around this time in 1904, Japan's victory over Russia, a major European power, marked the beginning of Japanese ascendancy in the Far East, and the first trigger in a chain of European upsets at the hands of Asian nations. In the aftermath of the Russian defeat, the two revolutions of 1905 and the Bolshevik revolution of 1917 successfully dismantled the 300-year-old Tsarist Empire. The fledgling Soviet Union, with its Eastern policy of supporting—some would say hegemonizing, especially after 1922—nationalist movements, exercised enormous influence over Asian anticolonialism. In India, after the Partition of Bengal in 1905, the rise of the *swadeshi* (national self-sufficiency) movement resonated with parallel developments elsewhere: with Japanese, Korean, as well as Chinese nationalisms. In China, following the revolution of 1911, and the May Fourth cultural movement of 1916, debates around Westernization and indigeneity paralleled the Indian discourse of *swadeshi*. Arguably this sequence continues till the end of the Second World War: in 1945, a different dispensation emerges with decolonization (first in Asia and then in Africa) and the Cold War between Soviet and American-led camps. Tagore's lecture on world literature, then, marks the beginning phase of a continent-wide upsurge.

The second text under discussion, Mao Zedong's 1942 Yenan Forum talks on literature and art, delivered during the height of the Chinese liberation campaign against the Japanese, marks the other end of this historical sequence. The Yenan lectures are a milestone in articulating the relationship between 'proletarian' art and 'nationalist' cultural tradition. Between Tagore and Mao, to put it modestly, we witness the consolidation of an *Asian* notion of the 'East' with a remarkable emphasis on the idea of culture as a site of emancipatory practice: simultaneously the Hegelian-Marxist categories of dialectical philosophy, previously restricted to elegant explanations of bourgeois Europe, encounter and have to grapple with a new, foreign and properly world-historical object of study. Non-European societies, those in Asia, Latin America, and to some extent

Africa, come to take centre-stage during this period, eclipsing and thoroughly reconfiguring the European conception of the world.

It might, still, seem odd to relate Tagore and Mao on aesthetics and literature. If that is so, that has more to do with the historical (or, extra-aesthetic) outcome of the twentieth century, and the latter's impact on the peculiar omissions and erasures—reinforced by disciplinary and institutional boundaries of knowledge—that characterize our present moment. Tagore and Mao represent two ends of a shared dialogue in India and China during the interwar era that was remarkable in putting forward a distinctive conceptual framework for modern literature and articulating the relationship between progressive aesthetics and politics. From the mid-nineteenth century onwards, the East India Company's export of opium from India to China, and the reverse trade in Chinese-made sugar to India, created a lasting legacy of British colonial penetration of Indian and Chinese markets as well as agriculture. Since early in his career, Tagore himself was aware of these issues and wrote on them intermittently; we learn that

> his articles 'China Maraner Byabasay' (The Death Trade in China) published in the Bengali magazine, *Bharati* in 1881; 'Samajbhed' (Social Differences) written in 1901; and the famous 'Chinemaner Chithi' (Letters of a Chinaman) based on [Goldsworthy Lowes] Dickinson's *Letters of John Chinaman* published in 1898 are eloquent evidence of his knowledge of and interest in Chinese affairs.
>
> (Das 96)

Here one finds discussions—based, admittedly, on thin sources and reports—of various topics, such as the agricultural crisis, Christian missionary activity, Chinese Confucianism, and symbolic art. These periodic pieces point to Tagore's growing engagement with China.

 This was not a one-way street. Chinese intellectuals' interest in Tagore grew significantly over time, especially after he won the Nobel Prize in Literature in 1913, the first Indian and Asian author to do so. Portions from his Nobel-winning book of songs, *Gitanjali*, were translated as early as 1915 by the Trotskyist intellectual Chen Duxiu (co-founder of the Chinese Communist Party, later expelled), and several well-known short stories, such as 'Chhuti' and 'Kabuliwala'; novels, such as *Ghare Baire* (The Home and the World); essays, and plays, were disseminated and widely read, both in English and in translation, throughout the 1920s. Notable May Fourth intellectuals such as Liang Shuming, Qu Quibai, Mao Dun, and Hu Shi, to name only a few, engaged with Tagore's work, sometimes admiringly and sometimes with criticism. The culminating point was Tagore's visit to China: in April and May 1924, he delivered a series of lectures to the Beijing Lecture Association, as well as in Nanjing and Shanghai. Published as *Talks in China*, the texts of these lectures constitute a crucial if ignored episode of modern Indo-Chinese intellectual history. While Tagore's China lectures are beyond the scope of this essay, his *visva-sahitya* talk delivered in Calcutta seventeen years earlier clearly illuminates the continuities in his core ideas on literature and human freedom.[1]

In *visva-sahitya*, Tagore freely articulates what we can call a parallel Asian vision to that of Vico and Hegel in Europe, namely an idea of 'socialist universalism'. The first term of this pairing is often obscured, and even ignored, in focusing on the second. But clearly, Tagore, in speculating on the scope of individual freedom, locates that freedom in the development of humanity-in-common; he "seek[s]," in the individual, "a union of its particular humanness with all humanity" (*samasta manusher maddhe sampurnarupey apnar manushatter milan*) (141).[2] A little-known aspect of Tagore is his decades-long and deep investment in the Bolshevik experiment; he visited the Soviet Union for a two-week trip in 1930, and his observations, published as a collection of *Letters from Russia*, evince an admiration of the new system quite common to anticolonial intellectuals of the era, even as he presciently noted the restrictions on freedom and worried what that might portend for the future development of that society. Among other things, Tagore praised the massive strides made in Soviet primary education, an initiative he saw as paralleling his own efforts in his native Shantiniketan, as well as the Soviet attempt to accommodate hitherto marginalized nationalities within the socialist system, which Tagore saw as a possible model for decolonization in British India and beyond. Tagore, it is true, was not a socialist in the strict sense or a fellow traveller of the Bolshevik regime; however, his empathetic understanding of the human condition under socialism and his philosophical belief in collective emancipation far exceed the limits of contemporary bourgeois-liberal universalism.

In Tagore's expansive view of subjective interrelations, one that is arguably resonant with Vichian-Hegelian dialectical philosophy, what is posited is a double connotation of the 'world'. On one hand, the world is the concrete manifestation of subjective intention and action; on the other hand, the truth about the objects of this world can be perceived, progressively, only through subjective understanding and experience. Here it is noteworthy that Tagore does not devalue subjective, or immediate, understanding. If subjective understanding is "incomplete," which it is—when it is "by itself," Tagore adds—that is simply because the human spirit (*atman*) is in the process of discovering and searching out reason, or "truth," which is a matter of historical development. The dialectical thrust of his argument once again deserves underscoring since he notes, "One can apprehend one's innermost spirit (*antaratman*) in the outside world most readily and comprehensively in other human beings . . . [t]he more I apprehend that unity, the greater my good and my joy" (139, 141).

Tagore's notion of this world as historically formed, as is the "truth" of the world, which is "fashioned by many people through many years" (142), alerts us to the materialist rather than idealist underpinning of his thought. Of course, he goes several steps further, as we will soon see. Included in historical development is sensory perception, such as "sight, hearing, and thought," "the play of the imagination," and "the attachment of the heart" (139), all of which contribute to the knowledge of the world as it is perceived, felt, and ultimately known. It is in relation to sensory perception that literature emerges in Tagore's discussion. Dividing the realm of human activity into work and literature, *karma* and *sahitya*, it is to the latter that Tagore assigns the crucial task of forming "affective bonds with truth," one which goes beyond self-centredness in seeking a universal realm of humanity.

At this point, the difficulty that arises in locating Tagore within a Marxist prism—a task that specialists have shied away from—is the seeming devaluation of work. Tagore *appears* to suggest that work is driven by instrumental, self-serving reason. It is also true that Tagore speaks of work as an "offshoot" of human expression (142), almost a mark of humanity's fallen state rather its true purpose. Furthermore, he appears to view literary expression much more positively as "the prodigal wing of human nature," and as that which repeatedly thwarts the designs of "reason, the parsimonious steward" (143). Yet it would be superficial to interpret this position as prioritizing either literary endeavour over laborious work, or favouring imagination over reason. Tagore is speaking in Vichian terms here, emphasizing the importance of oral myth and the popular roots of understanding, as well as the role of the literary imagination in actively shaping the material world. As I have already indicated, Tagore's method is dialectical, intent on finding relationalities rather than dualist antitheses.

In this essay, Tagore's genuinely astute insight is the suggestion that literature is at once the most intensely personal exploration of the world around us, driven by the self's urge to express itself, while at the same time, literary work seeks to overcome the "obstacles" created by "self-interest" (146). Literature and work, rather than being antithetical, Tagore notes, are "parallel" to and "complement each other" (142). What literature embodies, then, is a *special* kind of work. If the self is at its most free when it is able to express itself fully, such as in literary expression, such "total autonom[y]" also leads to "isolation," and to a state of— Tagore uses a strong word here—"uncivilized" existence. The growing freedoms of modernity result not only in a loss of authenticity and human essence, as post-Romantic philosophy in the West would hold. Even the 'civilization' fostered by such freedom is but the negation of the actual concept. Meanwhile, it is in literature that the self, in articulating, simultaneously fulfils and abolishes itself, thereby approaching 'totality' and a higher realization of the self. Nowhere in the essay does Tagore use the term *totality*. His preferred word for the process, from Hindu and Buddhist philosophical sources, is *samsara*, which is the cycle of rebirths that characterize the phenomenal world.

This is Tagore's kernel, then: while literature comes to serve as the expression of the universal spirit, what he terms "concrete form in the world" from "inchoate . . . abstract ideas," it does so through self-consciousness "reveal[ing]" "itself to itself" (142). Rather than rejecting the individual in the face of the general or the universal, literature testifies to the importance of individual self-consciousness in any manifestation of the universal spirit. As he says, "Incomplete in itself, it [individual consciousness] is relieved if it can somehow turn its inner truth into the truth of the world" (143). At the opposite end of the spectrum, the other possibility, as it were, is the greater alienation and stunting of the self (*oudasinya*), a resigned withdrawal from the world that Tagore compares to death (*mrityu*).

Tagore in my view offers a nuanced view of the role of literature in the context of (colonial) modernity. He is a philosopher of the modern, someone who insists, in the best traditions of dialectical thought, on the need to engage with the world as it *is*, without withdrawing from it in despair or taking recourse to nostalgia for the past. Georg Lukács' characterization of Tagore, on reading the latter's novel *The Home and the World*, as a metaphysical and reactionary thinker is, in this respect, worthy of revision. If anything, Tagore systematically challenges the irrational and nostalgic strands of anticolonial thought in India, and puts himself more in conversation with Left-Hegelianism than what Lukács' caustic review, titled 'Tagore's Gandhi Novel' (1922), sees in the Indian thinker.

A few additional comments are in order about the other salient points in his argument. In the *visva-sahitya* essay, Tagore defends the right to property, and more controversially the passing of property from father to son. But it is necessary to point out that he does so on the basis of socialized property rights, and insists rather radically that this applies to everyone rather than only to the landed elite. In a nation where hierarchies of caste, not to mention class, inhibited and continue to inhibit the rights of inheritance, this is not a small claim to make, although it is troublingly silent on the patrilineal aspect.

Similar too is Tagore's defence of war in *visva-sahitya*. Tagore sees modern warfare as the positive expression of the collective spirit of nations, yet he is equally quick to point out that such intentions have been subverted by— "Western"—nation-states that instrumentalize war through mechanical means (weapons) and for the accumulation of resources and profit. In other words, to paraphrase his contemporaries Vladimir Lenin and W.E.B. Du Bois but almost a decade before them, Tagore here is making a distinction between oppressor nationalisms and those of the oppressed (colonized nations), and emphasizing the right of the latter to self-determination by all means, including war.

Third and finally, by positing history as entirely human-made, he advances a view of modernity that is based on progress, denying an unreflexive admiration for the achievements of the past. While *visva-sahitya* points to the growing stratification and alienation that is a direct result of capitalist modernity and its instrumental use of reason (which, it is needless to say, would hardly appear otherwise to a colonial intellectual), Tagore insists on affirming the historical development of human reason (*buddhi*) and its complementary relation to affect and emotion (*hriday*).

Tagore is no garden-variety admirer of the modern. In fact, it is his critical unpacking of the pitfalls of modernity that most resonates with our own historical consciousness. He did not live to see the outcome of the twentieth century, but already in his time he described a globalized world marked by war, environmental destruction, vast immiserated populations, uprooted communities, and the isolation of the precariously free individual. As a special kind of labouring work, it is literature (*sahitya*) that offered to him a way of narrating—as well as negating—this world (*visva*). Such an articulation of the contradictory aspects of the modern material world, and the role of human activity in it, moreover, places Tagore in contradistinction to current versions of theory that promote anti-modernity and posthumanism as radical pathways. In the theoretical framework that Tagore put forward in *visva-sahitya*, modernity, rationality, and universality are processive notions to be achieved through human effort. It is closer in that sense to Marxian materialism.

Marx and Tagore were of course addressing distinct notions of the proletariat, as befitting their different socio-historical locations. Both, however, commonly understood the term to be referring to those who by virtue of their self-activity were singularly responsible for the reproduction and perpetuation of society, without necessarily having attained the consciousness of doing so. The dialectic between speculative thought and social being, and between transcendence and actualization, posed a properly universal, or ex-European, problematic only in the twentieth century, shared alike by socialist thinkers in Europe and anticolonial intellectuals in Asia. At the same time, in the third figure of Mao Zedong after Marx and Tagore, there emerges a new trajectory, and an astute reformulation of the relation of speculation to totality, one that is rooted in the specifics of twentieth-century anticolonial struggle and especially over definitions of 'the people'.

In *New Science*, Giambattista Vico observes in his well-known passage on 'Method' that in order to trace the development of human thinking:

> With the philosophers we must fetch it from the frogs of Epicurus, from the cicadas of Hobbes, from the simpletons of Grotius; from the men cast into this world without care or aid of God . . . Our treatment of it must start from the time these creatures began to think humanly . . . we ha[ve] to descend from these human and refined natures of ours to those quite wild and savage natures.
>
> (100, par. 338)

Vico seems to have been aware of the highly uneven aspect of popular consciousness, fuelled no doubt by the Neapolitan philosopher's understanding of eighteenth-century Italian city-states and their competing claims to sovereignty. This problematic of unevenness—what he terms the "national" component of *Geist* or Spirit—is taken up nearly a century later by Hegel in his *Lectures on the Philosophy of World History*, and even later in the twentieth century, by Vico's compatriot Gramsci.

On the other hand, Mao, who takes up a similar thematic in his Yenan talks, is responding to his contemporaries in the May Fourth movement, to the debate between the Westernizers and neo-traditionalists of various stripes, as much as to the European tradition. In the Yenan lectures, the methodological point that Mao makes about intellectual work is remarkably resonant with Vico's advice to the philosophers, namely that one has to make sense of creative imagination, what Vico terms "divinity"—to divine, in secular "gentile" fashion, is to imagine—from the material conditions in which the people operate. It is wrong, Mao holds, for intellectuals to look at the culture of the past and to deduce from them the character of the people; such an approach takes forms, which are but ideal projections, and substitutes them for actual social relationships.

Mao takes two seemingly antithetical positions at once. On one hand, intellectuals have to engage, critically, he says, with "the hangover of petty-bourgeois ideology among many proletarians and backward ideas" (4). On the other hand, he asserts that many intellectuals "were acquainted neither with their subjects nor with their public" (6). The former suggests a vanguardist notion of consciousness-raising *from above*; at the same time, much of what Mao says about the intellectuals' lack of familiarity undercuts and, in many ways, fundamentally alters the previous position to propose a different vision of literature and art. Just as Vico had, quixotically, proposed moving from Epicurus, Hobbes, and Grotius to undertake a study of the "frogs," the "cicadas," the "stones," and "rocks" to illuminate the "savage natures" of the early humans, and emphasized the role of the "philologians" in the undertaking, so too does Mao. Impatient with "abstract ideas," Mao claims: "In the life of the people there lies a mine of raw material for art and literature" (21). A fairly commonplace suggestion, that: what is distinctive is Mao's two-fold insistence that the "life of the people" be understood first and foremost through things traditionally considered by intellectuals to be "dirty" (7) and "crude" (21); a second aspect, following from this previous one, is a call for a historical-materialist philological enquiry, that "language" (6–7) itself be investigated afresh.

Mao places language, and philological enquiry, at the very heart of the "objective study" (10) of human society. He emphasizes that language is key both to the development of inner, psychic life, as well as to cultural expression; the latter includes oral and plastic art in addition to written texts of literature. Language not only registers but also dialectically determines human, proletarian existence, including "their [people's] emotions, their manners . . . [as well as] wall newspapers, murals, folk songs and folk tales" (16). Changing patterns of human interaction with the material world, and the historical evolution of modes of living, are traced in the field of language. Language, for Mao, is a dynamic (constantly changing) index of historical human experience. It needs to be pointed out that such a conception of language runs counter to the linguistic approach canonized in theory post-Saussure, where language is held to be a closed, rather than open, system, with no grounding in history. In fact, the very category of 'history', as something that is material and human-made, disappears in the poststructuralist redefinition of language; Mao's interlocutor in postwar France, the anti-humanist philosopher Louis Althusser, best exemplifies this paradoxical trend.

On the other hand, the Yenan lectures have been criticized, with good reason, for propagating a partisan, 'party line' view of artistic production. Such criticisms are well known and need not concern us here.[3] For his part, Mao refutes the orthodox conflation between politics and art, and in fact devotes the entire last two sections of the lectures to make the point that political and aesthetic criteria are differently based, not the same thing, and should not be substituted for each other. In short, artists and intellectuals enjoy a degree of autonomy separate from political activism; their concerns are also somewhat distinct from the prevailing milieu. A significant number of statements and passages from the lectures could be cited to support this view. What is more important, however, is his view of past traditions of national cultural forms. In the face of the traditionalist and neo-Confucian insistence that ancient culture contains the necessary wisdom for China's future growth, the Yenan lectures posit the past as 'foreign', something that is neither to be accepted or rejected wholesale but to be re-fashioned.

If Vico, in *New Science*, had presented the past as something that is truly comprehensible only to those who lived it, and therefore foreign to the present, Mao takes the Vichian position further. Echoing Vico, Mao argues that national character is historically uneven at any given moment, including the present; depending on levels of human self-actualization, different "historical epochs" co-exist within the same nation (45). Such unevenness further implies that cultural forms belonging, strictly speaking, to different historical eras are to be found co-existing with one another in the present moment. Mao proposes what is undoubtedly an *extension* of the Vichian philological project, namely the "use . . . [of] the artistic and literary forms of the past," and imbuing, "reshap[ing] them and filling them with new content" (14). He argues that the socialist intellectual cannot be blind to the past like the modernizers, or to the outside world like the nativists. The correct approach is immanent and relational, "not reject[ing] the legacy of the ancients and the foreigners, even though it is feudal or bourgeois," but combing these with "our own creative work" (22). Arguably, this is a theory of world literature where both the past and the non-national, *as foreign forms*, interact and intersect with the form and content of the local and immediate present. In other words, what is entailed is the practice of "constellation" (Benjamin), neither rejection of the foreign and the past nor their passive emulation, but activating them under new conditions so as to articulate previously unavailable truth.

Contrary to current assumptions, Mao's contribution to materialist thought does not solely rest on de-emphasizing the industrial proletariat of nineteenth-century European social democracy to include the 'postcolonial' category of the peasantry. In line with Tagore, Mao's theorization is invested in the continual dialectic of human self-consciousness actualizing itself, in the realm of language and within cultural forms. Such an emphasis does not simply pose the peasantry as another economic category to be reckoned with, especially in the non-European world, but resists economic determinism itself, which in the twentieth century assumed unquestioned prestige in Soviet Marxism (and, to be clear, in post-Maoist China too), fascism, and the liberal-capitalist models of Fordism and Taylorism. For this grouping, and the various imitative regimes of third-world nationalism, notwithstanding differences in ideology, the development of productive forces and rational management of economic crises were the solution to modernity's ills, leaving little to no philosophical room for purposive human agency. Today, we are far more sceptical of such methods. It is in this context that Mao's Yenan lectures must be judged, in pointing to the dialectical manoeuvre between intellectuals and cultural practitioners (rather than traditionalist ontology or scientific positivism), and their subject of work—social reality.

Unlike Tagore's lesser-known lecture, Mao's Yenan lectures on literature and art have a long institutional and cultural afterlife that is not limited to China alone. What stands out in this process of global circulation is the notion of humanist transformation. If 'Maoism' found a captive albeit ill-informed audience in the New Left circles of the West in the 1960s and 70s, as well as among activists of the third world, part of that appeal surely lay in the philosophical affirmation of the figure of the human, namely the interaction between transcendence of the existing order and the realization of speculative thought. In this mode, the European intellectual tradition founded by Vico and Hegel paradoxically found its way back to the West through Mao's Asian dialectics. This occurred too (it has to be said) in the face of the despair of much of postwar Western, especially French, anti-humanist theory. On the other hand, what is elided in the Western reception of Mao is the enormous impact that his formulations, especially on popular sovereignty, had on African and Latin American struggles against postcolonial nationalism and neocolonialism. In the second half of the twentieth century, no other Asian figure—including M.K. Gandhi—exercised comparable influence in the non-European world.

It is beyond the scope of the present essay to expound on these ideas, connections, and exchanges between the Western and postcolonial worlds, as conventionally labelled. In the spirit of Timothy Brennan's intervention, what remains to be recovered more fully, however, are the contributions of anticolonial intellectuals, male and female, in various parts of the erstwhile third world in the tradition of cultural materialism. As Brennan challenges us to consider, the thought of Giambattista Vico offers a unique and crucial perspective on cultural materialism, not simply as an analytical framework but as a form of resistance politics for transforming society. It is Vico who allows us to read language and literature as both marked by historical processes of global *class* struggle; similarly, his insistent return to the subjects of popular myth and orality, and critique of class and foreign domination, continues to offer templates for peripheral and semi-peripheral societies struggling with international sovereignty and grassroots democratic participation. Vichian Marxism offers a global perspective, in short, that speaks to the issues of our present.

Notes

1 At least one scholar, the Indian comparatist Sisir Kumar Das, has noted that, "Although *Talks in China* is one of the significant writings of Tagore in English, it never received the attention it deserved either from Tagore scholars in Bengal or students of India-China relations" (87).
2 All quotations in English are from Chakravorty's translation of the lecture (Tagore 'World Literature'). Where relevant, the Bengali original has been consulted (Tagore 'Visva-sahitya').
3 Liu Kang provides a fine account of China's revolutionary period and its aesthetic lessons for Western, post-revolutionary critical theory. As he points out, the Yenan lectures underwent many revisions in published editions, each of which reflected the party-state's attempt to gain control over past socialist legacy.

Works cited

Adorno, Theodor. *Minima Moralia: Reflections on a Damaged Life.* Translated by E.F.N. Jephcott, Verso, 2005.
—. "Resignation." Translated by Wes Blomster, *Telos*, vol. 1978, no. 35, Mar. 20, 1978, pp. 165–168.

Auerbach, Erich. "Philology and 'Weltliteratur.'" Translated by Maire Said and Edward Said, *The Centennial Review*, vol. 13, no. 1, Winter 1969, pp. 1–17.

Benjamin, Walter. *The Origin of German Tragic Drama.* Translated by John Osborne, Verso, 1998.

Brennan, Timothy. *Borrowed Light: Vico, Hegel, and the Colonies.* Stanford UP, 2014.

Casanova, Pascale. *The World Republic of Letters.* Translated by M.B. DeBevoise, Harvard UP, 2004.

Das, Sisir Kumar. "The Controversial Guest: Tagore in China." *India and China in the Colonial World*, edited by Madhavi Thampi et al., Social Science P, 2005, pp. 85–125.

Friedman, Susan. Planetary Modernisms: Provocations on Modernity Across Time. Columbia UP, 2015.

Hegel, Georg W.F. *Lecture on the Philosophy of World History.* Translated by H.B. Nisbet, Cambridge UP, 1975.

—. *Phenomenology of Spirit.* Translated by A.V. Miller, Oxford UP, 1979.

Kang, Liu. Aesthetics and Marxism: Chinese Aesthetic Marxists and Their Western Contemporaries. Duke UP, 2000.

Lukács, Georg. "Narrate or Describe?". *Writer and Critic: And Other Essays*, edited and translated by Arthur Kahn, Authors Guild, 2005, pp. 110–148.

—. "Tagore's Gandhi Novel." *Die rote Fahn*, 1922, marxists.org/archive/lukacs/works/1922/tagore.htm. Accessed Aug. 6, 2017.

López, Silvia. "Dialectical Criticism in the Provinces of the 'World Republic of Letters': The Primacy of the Object in the Work of Roberto Schwarz." *A Contracorriente*, vol. 9, no. 1, Fall 2011, pp. 69–88.

Majumder, Auritro. "The Case for Peripheral Aesthetics: Fredric Jameson, the World-System and Cultures of Emancipation." *Interventions*, vol. 19, no. 6, pp. 781–796.

Moretti, Franco. "Conjectures on World Literature." *New Left Review*, 1, Jan.–Feb. 2000, pp. 54–68.

Said, Edward. *Culture and Imperialism.* Vintage, 1994.

Schwarz, Roberto. *Misplaced Ideas: Essays on Brazilian Culture.* Edited and introduced by John Gledson, Verso, 1992.

Spivak, Gayatri. *Death of a Discipline.* Columbia UP, 2003.

Tagore, Rabindranath. *Letters from Russia.* Visvabharati, 1984.

—. *Talks in China.* Rupa & Co., 2002.

—. "Visva-sahitya". *Rabindra-racanabali*, vol. 8, Visvabharati, 1941, *Bichitra: Online Tagore Variorum*, bichitra.jdvu.ac.in/index.php. Accessed Aug. 6, 2017.

—. "World Literature." Translated by Swapan Chakravorty. *Selected Writings on Literature and Language*, edited by Sukanta Chaudhuri, Oxford UP, 2001, pp. 138–150.

Thiong'o, Ngũgĩ wa et al. "On the Abolition of the English Department." *The Postcolonial Studies Reader*, edited by Bill Ashcroft et al., Routledge, 1995, pp. 438–442.

Vico, Giambattista. *The New Science of Giambattista Vico.* Translated by Thomas G. Bergin and Max H. Fisch, Cornell UP, 1984.

Williams, Raymond. *Marxism and Literature.* Oxford UP, 1977.

Zedong, Mao. Talks at the Yenan Forum on Art and Literature. Foreign Language P, 1960.

2 "Le mot du poète, le mot primitif"

Aimé Césaire and Vico's civic humanism

Eric Brandom

Timothy Brennan's *Borrowed Light* makes a powerful case for investigating an intellectual history of emancipatory and anticolonial theory with Giambattista Vico and G.W.F. Hegel as axial figures. A reconfigured, redeemed humanism would be central to this history. I argue here that Aimé Césaire is an important and usefully uncomfortable test case for Brennan's approach. This chapter looks to Césaire's earlier work, which is to say up to 1944, including his best-known poem, the *Cahier d'un retour au pays natal*. The French imperial nation-state, as Gary Wilder has argued, articulated national belonging, universalist culture, class, and race together in a powerful and powerfully disarming way. The problems presented by universalist republicanism, by the forms of humanism that under-wrote it, pointed Césaire in a direction that rewards comparison not only with arguments made 200 years earlier by the Neapolitan Vico, but also with thematics highlighted by Brennan's book.

René Ménil, writing in 1963 for a communist journal, in a well-known critique of Césaire drew a line between "black poetry" and Césaire's famous neologism, "*négritude*." The former was essentially good, the latter bad. The former, in expressing the life of colonized Blacks [Noirs], at the same time expresses the historical and social condition of blacks [nègres] in modern civilization . . . It seeks to describe the wealth of what is called the black soul [l'âme noir], which is to say the hates and loves, resentments and hopes, of the colonized black man [l'homme colonisé noir].[1]

The *Cahier* was an example of black poetry, while already in *Tropiques*, the journal Césaire had co-founded with Ménil in Martinique on returning from France two decades earlier, Ménil claimed, Césaire "glorified an anti-intellectualism and a superstitious mysticism" themselves drawn from "reactionary philosophical sources" (Ménil 38). For Ménil, Negritude and its antecedents possessed only a "pessimistic and violent critique of capitalism," rather than a Marxist one (Ménil 44). Edward Said, Brennan writes, in his work on the later nineteenth and early twentieth century, "meant to unearth traditions of humanism and rationalism in modernity that have been buried under modernism's cultural pessimism, irrationalism, and third-worldist invocations of the primitive" (Brennan 69). This chapter, then, investigates Césaire along the lines Brennan suggests—no more than partially, focusing on the earlier writings and raising rather than answering questions.

Just because of his perhaps anomalous stand against national independence in the immediate postwar, and his evident affinities and connections to the modernist avant-gardism of the Surrealists, Césaire is especially worth considering in terms of the anticolonial Vichianism Brennan limns out.

This chapter places Césaire almost entirely in terms of European traditions and interlocutors and therefore runs the risk of turning him into simply an exemplary European or, worse, Frenchman. He is one of the great French poets of the twentieth century—but he is not only that. Césaire reported in the 1970s that as a student in interwar Paris, "I was under the same influences as were all the more or less cultivated *French* students of this time" (Césaire and Ménil viii). Yet of course not all the cultivated French students met and befriended Léopold Sédar Senghor. Césaire belongs to the Caribbean, indeed to a Francophone and Black Atlantic that dates back at least to the cultural problematics that emerged in the decades after Haitian independence (Daut; Gilroy; Edwards). Indeed it is significant that he saw no option but to write in French, and that he was and remained so deeply invested in French and classical culture, because he was also a racialized subject, and at first only allowed to participate in metropolitan culture. The diasporic foundations of Césaire's situation are not to be minimized; nor, for instance, is the encounter with an African culture, defined for him to a great degree by the German ethnographer Leo Frobenius. In addition to Senghor and other French colonial subjects, we ought to bear in mind the (now lost) thesis on African-American literature that Césaire wrote in the 1930s. Claude Mackay and Langston Hughes were thus models, together with Renée Marin and Apollinaire. Yet Césaire for his own reasons and with different resources constructed an insurgent and disruptive humanism—a materialist and civic humanism—that bears striking similarities to what Brennan finds in Vico. To begin, then, let us turn to Vico.[2]

Reading Vico

The third and final version of Vico's major work, *New Science*, appeared in Naples in the year of its author's death, 1744. The book presented a cyclical theory of history according to which all gentile nations—those without the benefit of direct revelation—pass through a sequence moving from barbarism, through heroism, civilization, by means of excessive refinement to "civilized barbarism," and then returning to barbarism itself. These are his famous *corsi e ricorsi*. In the twentieth century, the most attention has been paid to Vico's *verum-factum* principle—"the true and the made are convertible"[3]—and subsequent distinction between the true and the certain (*verum/certum*). This is an anti-Cartesian idea. Descartes and his school completely dominated Naples by Vico's time, or at least so he felt, and so Vico was explicitly and historically as well as conceptually anti-Cartesian. It is only possible to know—as *verum*—what you have yourself made, the *factum*. Thus we can know the principles of geometry because we can and do *make* them again and again. On the other hand, God made the natural world and therefore is the only one who can know it. The cultural world was made by human beings and therefore can in principle be understood by human beings.[4] It is very difficult, requiring arduous scholarly effort and the active use of our *fantasia*, or imagination, but it is possible. Vico uses the *corsi* as an interpretive grille or canon to arrive at new results about the history of the institutions that have given rise to the cultural evidence he is now able to array in its correct order. One central text was the body of Roman law, which he regarded as a "serious poem." Although he did not use the terms in this way, Vico defended history against science.

He practised what might be described as a critical philology in which the cultural and linguistic production of a given society was read backward to the material conditions of its institutions. How we live—in a primeval forest, isolated dwellings, towns, crowded cities—determines our language, which determines our thought. Even our bodies participate in this history. The giants that humans were in the distant past still inhabit our language. At the centre of this approach to the cultural world is the human capacity to create and to imagine: What one human mind has created, another human mind may, in principle, subsequently imagine. Although of course Vico claims that this principle is really an empirical finding, the result of his research rather than a presupposition of it, it is axiomatic. It enables and is enabled by the *corsi e ricorsi*.

The universalist humanism of Vico's method cuts against what one might reject as his stagist, cyclical theory of history. Culture, which is in important ways coextensive with the material world we have built up around ourselves, is for Vico fundamentally open, a space that unifies us. Here is a materialism that cannot issue in racism because it presupposes the translatability of cultural productions. Yet against that backdrop of cognitive availability, Vico went against the Cartesian spirit of his time by insisting on the historical relativity of even the basic structure of human cognition. As Auerbach put it, "in Vico's system the old contrast of natural against positive law, of *physis* against *thesis*, of original nature against human institutions, becomes meaningless" (194).[5] So the human and the non-human are rigorously separate as far as interpretation goes, but by the same token one cannot bracket anything 'merely natural' out of historical human reality. Our bodies themselves are historical phenomena.[6] A basic translatability of different kinds of human being is axiomatic—it is a matter of faith. For the historian Jules Michelet, who claimed in the early nineteenth century to have been much influenced by Vico, it would be a political and ethical first principle. Descartes had explicitly rejected knowledge that might depend on the body because for him dependence in general was epistemologically disqualifying. Yet by the later nineteenth century the profoundly Cartesian French intellectual establishment had managed to bring bodies and their hierarchies back into the picture.[7] The racial hierarchies baked into the practice of French culture in the nineteenth century, theorized or denied, are an essential condition for Césaire's work. In fact, he arrived at positions akin to Vico's in order to criticize this Cartesianism.

Brennan foregrounds Vico as a resource for and goad to an anticolonial and philologically oriented humanism, a "civic hermeneutics" that places emphasis on "the vulgate rather than the classical; on secular and corporeal solidarities rather than sacred textual encounters; and on the circulation of demotic and experimental forms rather than their containment within notions of aesthetic autonomy" (Brennan 4). Vico articulates an early and influential version of this kind of humanism, and Brennan constructs, or indicates, a line running from the diminutive Neapolitan through Hegel above all, but also Herder, Lukács, and Said. At least as important is the cast of contrasting figures, most importantly Spinoza and Nietzsche, but also Georges Bataille.

Brennan's intellectual history is also a call for a certain kind of literary criticism. Looking to Vico, Brennan identifies some tendencies or proclivities of a Vichian "civic materialism based on the beauty of the tactile word." For Vico, according to Brennan, "words . . . are originally bodies, not signs" (39). Vico's theory of the historical development of language gave a privileged place to poetry and to the *spoken*, hence to orality. More broadly, Brennan argues that Vico "returns language to a profane illumination of quotidian objects of exchange . . . tending to reground and concretize along sociological lines the political aspirations of early twentieth-century avant-garde poetry" (41). Art can be political because "fledgling civil institutions took shape in poetic thought, and economics was an exchange of word-things" (41). The primitive, rather than a spectacle of otherness, is an always already present aspect of our own actions.

Vico's afterlife unfolds in interesting places.[8] Michelet discovered him in the 1820s and made the most influential translation into French, although not the one read by Karl Marx.[9] In the 1890s, Georges Sorel argued over the significance of Vico for Marxism with Paul Lafargue, son-in-law to Marx; in the 1910s, Vico was among the points of conflict between Benedetto Croce and his friend and rival Giovanni Gentile, philosopher of fascism. Vico articulated a non-Cartesian universalism and pushed some of his readers to give a special place in political thought to creative, poetic activity. That is, Vico has been a spur to and assistant for cultural politics. In the nineteenth century, this was of course a position also held by a certain romantic strain of philosophy. It certainly proved easy to draw on Vico in support of anti-modern projects.[10] Yet Vico stands apart from most of his nineteenth-century readers because he has provided an anthropological universalism that is at once materialist, cultural, and at least potentially anti-nationalist.

Vico represented, at least for Sorel and Lafargue, an approach to the study of culture that was an especially effective disruption to especially French ways of connecting political society, culture, and national belonging. Lafargue and Sorel both read Vico against the dominant culture and cultural politics of the Third Republic. For both, Vico served to broaden the materialism of Second International Marxism. For Lafargue, this meant critiquing the idealism of the time through a corporeally oriented study of symbolism. For Sorel, it meant regarding the productive process as one that exceeded capitalism not only as a social dynamic but also as institutional practice legible through culture. Sorel used Vico to try to read worker organizations as institutions with their own dynamic and meaning, spaces in which the postcapitalist (but, for Sorel, still productive and industrial) future has already begun to exist. He and Lafargue rejected the universalist humanism of the bourgeois-republican order, but sought to replace it with a new materialist and proletarian universalist humanism. Césaire, on the far side of the Great War, faced the problem of racism in a way that neither Sorel nor even Lafargue could have understood, and also much more immediately the possible dissolution of empire. They nonetheless had some of the same goals and opponents.[11] As a response to the racial coding of republican universalism, Césaire sought a universalism that would not pretend to subtract the racialized body. At the beginning of his career a poet rather than a social theorist or propagandist, Césaire tried to fashion a language and mythic framework out of his own situation that would make possible the kind of politics that he believed was demanded by Martinique in particular and, in the title of one periodical he would have read as a student, *le monde noir* more generally.

Culture in the French imperial nation-state

A significant body of scholarship on the post-emancipation period in the Caribbean has shown how conflicts over labour and autonomy for the formerly enslaved peoples tended to be resolved through racialization. Thomas Holt's classic study of post-emancipation Jamaica and the larger British Empire, for instance, showed how that imperial administration, liberal though it was, displaced political and economic conflict onto racial difference. In her study of citizenship in the Antilles, Silyane Larcher shows that these "old colonies" were and perhaps still are caught in a special corner of the Republican conceptual world. If full political rights appeared possible only for a moment after the 1848 abolition—the Second Empire in the 1850s abrogated or rendered meaningless many such rights—the principle that the rights assignable to a given territory could be made dependent on its history, even within the regime of legal equality of the Republic, remained. Disagreement over the nature and meaning of the Republic is arguably the defining feature of French political thought, but there was specificity to the form these debates took in the Antilles. Because of the state of exception and delay that existed in Martinique, citizenship meant confrontation over the actualization of rights shared between officially equal citizens—in these circumstances, citizenship is a "polemical phenomenon" more than a practice (Larcher 267).

This long history of citizenship in the Antilles is essential to understanding the form of the problems that Césaire set out to resolve or that, in any case, he had no choice but to live with. Larcher argues that the functional conceptual political order of French Republicanism "politicizes the past of societies and the socio-anthropological heritage of individuals issuing from them, which is a logic of racialization" (329). Césaire in his poetry set out to re- or differently politicize the socio-anthropological heritage of Martinique and by doing so to make race mean something other than what it did. Larcher writes that the political grammar of the post-slavery Antilles was characterized by "essentialist differentialism expressed in terms of universalism" (330). It should not be surprising that Césaire was also working very much within the same framework—that he sought to work out a differentialist essentialism against a false universalism. In sum, we might say, when the metropole objected that the historical and cultural aftermath of slavery precluded the full exercise of citizenship in the Antilles, in fact this was a racialization. When Césaire invokes "our complex biological reality" (1330), which sounds like race, what this in fact meant was historical and cultural inheritance: "is it not our task to achieve our total humanity?" (1330).

A. James Arnold emphasized long ago the essential ambiguity of Césaire's most famous neologism. Is Negritude "essentialism or historicism?"[12] For Arnold this is the choice between a position that valorizes a racialized difference and one that rejects racialization in favour of historical-social being. Larcher's account, among others, suggests that in the French context history has been a perfectly serviceable substitute for race. Gregson Davis has argued that the central movement of the *Cahier* is the discovery or fashioning of the *moi*. Agreeing with Davis and Arnold, but taking Larcher to provide a backdrop for Césaire's politics, we might look at things differently. The ambiguity that Arnold identifies is constitutive of the situation in which Césaire finds himself, and in fact both readings of Negritude—like the *moi* constructed and worked over the course of the *Cahier*—must be understood as rhetorical strategies in the construction of a political subject on the basis of the Republican and colonial situation of Martinique, of the Antilles altogether.

Situating Aimé Césaire

Aimé Césaire was born in Martinique in 1913. His father, a schoolteacher, ensured what Davis calls "early bonding . . . with the French language" (5). The whole family was committed to education, and moved so that Aimé could continue his studies at the Lycée Schoelcher, where he won a scholarship to continue study in the metropole. He arrived at the Lycée Louis-le-Grand in 1931, and enrolled at the École Normale Supérieure in 1935. In 1939, partly through the intervention of one of his teachers at the ENS, the first version of the *Cahier d'un retour au pays natal* appeared in the magazine *Volontés*, alongside other avant-garde poetry of the time, much of it directly topical. The poem was the product of intense concentration and labour, which he blamed in part for failing the agrégation exam. Césaire returned home with offprints of his poem, which had appeared only weeks earlier, in his suitcase. The Second World War began, and before long elements of the French military loyal to the Vichy government were in control of Martinique. Whatever else he might have wanted, his politics at least until the end of the war could really only be a cultural politics.

The *Cahier* thus emerged from the Paris of the 1930s, and spoke both to it and to Martinique. So what was Césaire's experience of this tumultuous place and time? How can this experience help us to read the *Cahier*, and what can it tell us about Césaire's subsequent activity in Martinique? An important point to make here is that already in 1935, however familiar he was or was not with major texts of the Marxist tradition, Césaire worked in an environment shaped by communism.[13] Michel Goebel has recently examined the ways in which Paris in the interwar functioned as a centre of diffusion for communism and anti-imperialism among a generation of students from around the world. But this diffusion was uneven and not without limits and contradictions. In a sense, the French Communist Party (PCF) had been co-founded by colonial subjects, but by the later 1920s electoral politics had begun to shape the party's priorities. The PCF was, according to Goebel, "unable to recruit large numbers of *indigènes* in the metropole, let alone overseas, or unwilling to provide the resources necessary" to do so (186). Nonetheless, the PCF was the only significant force on the French Left that in fact officially and practically rejected imperialism. And it did score some symbolic successes, for instance in 1935 electing to a municipal council in Épinay, Félix Marlin "Martinican transport worker and member of the Union of Negro Workers" (Goebel 185). Indeed Antilleans in Paris, partly because as citizens they could vote and tended to have higher levels of education, contributed significantly to the practical operations of the PCF—although "many of them [were] lawyers or other liberal professionals rather than workers" (Goebel 186). So although Antillean students like Césaire moved in a Parisian social space shared with colonial or semi-colonial subjects, they occupied a relatively privileged position within it—yet still were black in Paris.

In his early years at the ENS, at least if his involvement in the journal *l'Étudiant noir* is any indication, Césaire struggled to stake out an independent position that would, retroactively, come to be designated as Negritude. In the March 1935 issue is a polemic against assimilationism organized around the highly charged vocabulary of the "métèque."[14] Césaire, in good normalien fashion, puts a quotation from Jules Michelet, avatar of French historiography, at the top of his text: "the hard thing is not to rise, but in rising, to remain one's self" (1292)—we can read the whole piece as a commentary on it.[15] The line comes from the preface to Michelet's *Le peuple*, one of his best-known and most 'democratic' books. It is a portrait of the French people, their suffering and relations with one another. In the preface, Michelet describes his own humble origins. The real name of modern man, he writes, is "worker." Like the people, Michelet has worked, in fact as a typesetter: "before making books, I *composed* them physically; I assembled letters before assembling ideas, the melancholies of the workshop, the boredom of long hours, are not unknown to me" (Michelet i-ii). Elsewhere in the preface, Michelet describes his approach to writing history as resurrection, and even mentions Vico, as well as accepting for himself and for the democratic experience he represents the label of "barbarian." Certainly Michelet's image of new sap rising would have been congenial to Césaire (Michelet xxix).

In the same short text, Césaire attacks the assimilationist logic of imitation, of "l'Identique" but also, interestingly, of theatricality: "Black youth does not want to play any part, it wants to be itself" (1293). Speaking to the "jeunesse," Césaire nonetheless insists that to be a man—as opposed to a child—is to "walk without guide on the great highways of thought. Subservience and assimilation are alike: they are two forms of passivity." In contrast, "emancipation is . . . action and creation" (1293). And not just any action and creation, but a contribution to "universal life, to the humanization of humanity"—but such contribution requires that one remain one's self, neither imitating the other nor allowing obscurantism or indifference to prevent struggle.

These themes remain in the next text Césaire produced for *l'Étudiant noir*, "Conscience raciale et revolution sociale."[16] For the epigraph, Paul Nizan on materialism has replaced Michelet on social climbing. Not manhood, but revolution, is now the goal. Assimilation is again the enemy, although now the danger is, at least partly, to assimilate to a revolution that is not one's own. Having cited Dostoevsky, Chesterton, and Charles Dickens, Césaire attacks Paul Morand's 1928 novel-travelogue about Haiti, *Magie noire*. Rather than object to the stereotypical depictions of African-descended people in it, Césaire writes about Morand's "hero, 'assimilated' Occide" who "is also revolutionary: thanks to him, Haiti has its Soviets, Port-au-Prince becomes Octobreville; and much good may it do, if he remains prisoner of the whites, a sterile imitating ape!" (1299). Here, again, Césaire links the capacity for creative activity to political autonomy. Hence the importance of arriving at the conditions—racial consciousness—for this activity so that, as Césaire wrote, the Antillean could arrive at the great crossroads of humanity with something of his own, and be other than a "plaything." And certainly it is a form of cultural politics against capital when Césaire writes,

> if it is true that the revolutionary philosopher is the one who elaborates techniques of liberation . . . must we not denounce the stupefying culture of identification and place, under the prisons built up for us by white capitalism, each of our racial values like so many liberating bombs?
>
> (1298)

Both Lafargue and Sorel had also recognized the exceptionally powerful attractions of the dominant universalizing French bourgeois culture. The problem was more difficult for Césaire, but his solution—shaped by the intervening generations of Hegelianism and existentialism—is, like theirs, to read the universal against itself by way of the primitive, this later understood as essentially *radical*. One can neither simply assimilate to French bourgeois culture—put on the melon hat and pretend one has always worn it—or to the PCF, which after all really made room only for the most bourgeois and professionalized Antilleans, precisely the ones from whom Césaire most wished to dissociate himself.

In this brief early essay what we see is Césaire beginning to argue that racialized difference constitutes a structure of domination as fundamental as class difference. The form of this appeal to authenticity-in-revolt is crucial: "we must not be revolutionaries who happen to be black [noir], but properly revolutionary blacks [nègres], and the accent must be on the substantive as well as the qualifier" (1299). This is the task of a cultural politics that is also a materialist politics. Returning to the language of Nizan, Césaire writes,

> let us set ourselves bravely to our cultural work, without fear of falling into bourgeois idealism, the idealist being one who considers the idea as daughter of the Idea and as parent to ideas, when we ourselves see there a promise that cannot fail to open into ramifying acts.
>
> (1300)

If in 1935 Césaire is already working within a generally *marxisant* context, he is also very clearly trying to hold together specificity and universality; indeed he believes that it is only by accepting and deepening one's own specificity that one can truly be universal. As for Herder, for Césaire already in 1935, the *étudiant noir* can be fully human only by embracing his (more on that masculinity below) blackness—Negritude.

It is worth highlighting in this connection Césaire's early interest in Charles Péguy, poet of Joan of Arc and the French nation, killed in action in 1914.[17] In a text meant in some way to introduce Péguy to Martinique, he finds that poet's significance in "Civic heroism . . . and much more supple than is sometimes said, but there is also in Péguy another heroism, exactly the one that [Julien] Benda refuses him, heroism of the *clerc*. His was mysticism of truth" (1307). Péguy was a model—to be followed with a difference—for how an intellectual ought to comport himself, to *make* culture happen in a dissident way. Péguy's great accomplishments included, after all, not just the poetry and the personality, but also his *Cahiers de la quinzaine*, a periodical of the highest quality (from the paper and typesetting to the ideas therein). *Tropiques* should perhaps be seen as made in the model of Péguy's *Cahiers*, mixing the literary, scientific, cultural, and topical.

Césaire, of course, first made a *Cahier* all his own. In Césaire's *Cahier* there is obviously something of Péguy's civic heroism, of Michelet's historiography of identification, his acceptance of the label of barbarian (the "cruautés cannibales")—but also, of course, as has always been most confounding for readers of Césaire, there is a humanist challenge to French civilization's falsely universalist humanism and to the colonialism undertaken on the basis of this false humanism. This poem, Césaire's masterpiece, can certainly not be discussed in full here. But a few themes and movements are suggestive given what has been said above and in light of the Vichian position highlighted by Brennan. For instance, the movement of the poem is on the face of it an internal and lyric drama of self-discovery and re-making. It is, if you like, a phenomenology of a return home that is also the painful extirpation and re-working on the poet's own body of what has been grafted into it by French culture. We can put this differently: in the poem, the knowing and the making—discovering and re-making—are indistinguishable from one another. The subject being constituted in this movement is the subject capable of revolutionary action, making itself out of and finally in turn master of the European references that dominate Césaire's earlier writings. This is not an individual subject at all, but a collective one.

This text, difficult and obscure to begin with, also evolved significantly over time. Césaire wrote the first version of the *Cahier* in 1938–1939, and this version of the poem has a clearer structure (four parts: three equal and a final shorter one).[18] The language echoes Baudelaire and to some degree Péguy whereas later versions in 1947 and eventually 1956 will find inspiration in the aggressive tendencies of the Surrealists. Arnold comments, for instance, that the quest of the "Je" shifts—in 1939 the accent is on the "spirituelle," while by 1947 it is on the "militantisme sociopolitique" (138).[19] Confining myself here for the most part to the earlier version, I want to point out the degree to which, from a Vichian perspective, there indeed even here is plenty of socio-political militantism.

Physical geography functions in the poem to mediate between the individual and the unachieved, unfinished, degraded, social. Indeed in the 1939 version of the text, these are the terms of the opening strophes: "les Antilles," bays and coves, mud and dust; hunger, disease, and alcohol; the city ("cette ville"); all related to one another through the movement of the earth, light, and non-relation, since both the Antilles and the city are "échouées," which is to say wrecked, washed-up, failed. Later versions of the poem, beginning with Bordas in 1947, have a new opening strophe, which has a very different set of referents and internal relations. Here is an "I," vituperative, speaking to a "you," invoked to reject law and order, hope, disaster, and the African landscape. In 1939, the mode is declarative rather than vocative, and the first human subject we encounter is either the abstraction of "les Antilles" or, more plausibly, in the third strophe, "l'affreuse inanité de notre raison d'être"—defined by negation and much less stable than the later version's given speaking subject. Martinique appears in the poem as a society in which none of the French ways of organizing national belonging function. Public monuments—from Josephine, wife of Napoleon who brought slavery *back* to Martinique after the Revolution, to Schoelcher, whose legislation abolished it in 1848—are without effect. Neither the schoolteacher nor the priest can solicit reaction from the hungry child (15).[20]

The poem famously begins with "au bout de petit matin . . .". The next few strophes, however, begin with "cette ville inerte." The city is also flat, corrupt in its blood—typified by mud, horse-trading, prostitution, alcohol, debauchery. The failure of the city to reach up into the sky—like a woman, it remains flat—is damning. It "rampe sur les mains sans jamais aucune envie de vriller le ciel d'une stature de protestation" (23). Spatial organization is actually a gender-coded corporeal one (male verticality vs female horizontality). In strophe 34, Haiti "où la négritude se mit debout pour la première fois et dit qu'elle croyait à son humanité" is juxtaposed with "la comique petite queue de la Floride où d'un nègre s'achève la strangulation." And from this corporeal geography, from a litany of slaving cities and slave territories (including US states), the red earth—sanguinaire, consanguine—we come to "Ce qui est à moi aussi: une petite cellule dans le Jura," referring to the prison in which Toussaint L'Ouverture, locked up by Napoleon, finally died. It is significant, given Césaire's political programme, that we approach Toussaint himself, the individual who as much as any other stands for Haiti, through the rock and snows of Europe.[21] And Toussaint unlocks, as it were, the first appearance of what would be a common technique in Césaire's poetry, a series of short lines beginning with the same word or words, here, "c'est un homme seul" and then, "la mort."

Collectives appear in the poem as essentially material, framed in by something, while the speaker of the poem, usually fully embodied, manages to transform into pure will. Already in the 1939 version of the text, a key role is played not only by cities—always failed in some way—but also the slave ship. The latter has of course deep significance as the topos both of social death and, as in Césaire's poem, the rebirth of a new collective subject. The extraordinary sequence of strophes 48–52 begins with an image of a city overrun by horse-traders, passes through the well-known passages rejecting the idea of some kind of glorious ancestral past, but also asserting the existence of a race "unknown to the chancelleries," defying the skull-measurers, which is that race constituted by the historical experience of slavery, and passes directly to the long strophe staging an encounter, in a Parisian streetcar, during which the poet laughs with two white women at a "comic and ugly" black. That encounter has been read as a moment of recognition of class difference, and it is, but it is also an evocation of a transversal solidarity, a city in potential in the interstices of the imperial metropolis. In these passages, at the centre of the poem, Césaire explicitly constructs an identity that is historical, although no less physical for that, against the political and scientific rationality of Europe (chancelleries, craniometers, and streetcars). Indeed in these passages the historical, geo-physical, corporeal, and civic are drawn together with extraordinary density. The *moi* here is never simply an individual.

In *The French Imperial Nation-State*, Gary Wilder sees the first parts of the poem especially as staging the problem of speech itself, but finds the poem unable to adequately "work through" the problems that it has itself posed. For Wilder the poem asks "what mode of discourse would allow the poet to avoid colonial complicity?" (283). According to Wilder, "refusing to choose between rationality and irrationality, myth and order, the poem displaces the very oppositions between rationality and irrationality, knowledge and myth, upon which colonial order was grounded" (284). Recognizing that Césaire is working a cultural and political problem, Wilder nonetheless, perhaps because he has begun with the question of voice, takes the *moi* of the poem as not univocal but as finally an individual. Indeed it is not clear why complicity—the logic of guilt and morality—ought to dominate the political here simply because we are in the mode of lyric poetry. Thus, "rather than celebrate the realization of authentic racial self-consciousness (for which the poem would become famous), the text in fact traces the fragmentation of racialized subjectivity" (286). If we are dealing with the subjectivity of an individual, then indeed we might agree with Wilder that

> there is no longer any trace here of the poem's earlier critique of voice and Negritude as necessary but inadequate . . . the concluding verses . . . provide a decontextualized and existentialist account of unalienated identity and metaphysical arrival. Natural and numinous metaphors displace historical and political ones.
>
> (288)

Through Vichian eyes, however, natural metaphors are certainly also historical and political. For instance, the description of the "Je" in a "pirogue" or canoe, praying, as it were, for tranquillity and transcendence, strikes Wilder as a failure, a decontextualized act of sheer mystical will, a retreat to subjectivity as a solution to history. We need not read the "pirogue," or the end of the poem, in this way. The image—a poetic image—yokes together through the Atlantic at once Africa, the Caribbean, and of course Paris itself—*fluctuat nec mergitur*. For Wilder, however, "this is poetry as prophecy. No longer a political instrument, voice becomes a sacred power outside and above phenomenal reality . . . Triumphant Negritude and cosmological universalism converge in a transcendental phallic fantasy" (288–289). The poem certainly is doing these things. But that is the point, it *is doing*. The speaker of the poem is not the same as the poet. At the end of the poem, we have passed out of the constative and into the imperative.

Wilder glosses the "return" to the "black hole" at the end of the poem as a Nietzschean as much as a Hegelian moment of self-founding individuality. For Wilder,

> Césaire's attempt to overcome the antinomy between universality and particularity ultimately reproduced it . . . the *Cahier* . . . concludes with a dual assertion of radical particularism and cosmological universalism . . . But the relationship between these unintegrated positions is never adequately worked through.
>
> (290)

In what sense can a poem work through this dilemma? How would that look? Wilder himself points out that "racialized subject-citizens attempting to participate in republican civil society confronted an overdetermined structural dilemma, not an intellectual puzzle" (201). We might respond by suggesting that what matters is how the poem's end in 1939 became the project of *Tropiques*, itself understood very much as a beginning of a new universality, an act of *collective* will and creative energy but also a new grounding and attachment to tradition. Not a modernist origin, but a beginning in Said's sense of the term.

We ought to return here to the problem of theatricality, of play-acting, of presentation, which we saw was a target of Césaire's ire already in 1935. Is literary language mimetic? Is it creative? It is fatal to play a role handed to you by the colonizer—but are all forms of theatrical self-presentation to be rejected? What if it is a role you have taken and made your own, rather than one that has been given? Who is the agent and who the actor in such a production?

Césaire, in the first part of the poem, introduces touches of naturalism. We might be in a Zola novel, reading moral corruption off the squalor of the family home. Still, the poetry remains hermetic in its vocabulary and in the tension put on the syntax of the French language. At the end of what Arnold identifies as the first movement of the poem, there is Césaire's extraordinary admonition against mere spectatorship or voyeurism, certainly an essential element of Zola's naturalism. The narrator describes leaving and returning, and then speaking, first to the "pays," and then,

> Et venant je me dirais à moi-même:
>
> « Et surtout mon corps aussi bien que mon âme gardez-vous de vous croiser les bras en l'attitude stérile du spectateur, car la vie n'est pas un spectacle, car une mer de douleurs n'est pas un proscenium, car un homme qui crie n'est pas un ours qui danse . . . »
>
> Et voici que je suis venu!
>
> (30)

We might put Wilder's argument that the poem problematizes voice in a different way: the poem rejects any idea of the literary as distinct from the world, of the body as distinct from the soul. That this eloquent attack on theatricality is itself staged within the poem, though, suggests that this untenable position nonetheless cannot be abandoned. The cultural, the literary, can only take us so far. A contradiction, then, highlighted and in this way pointing beyond the poem. Perhaps we can read here the beginnings of Césaire's turn to drama, which indeed he was already considering in the middle 1940s ('Ur-texte' in Cheymol and Ollé-Laprune).

The presentation of the first issue of *Tropiques* highlights the problematic of man. We start with the "mute and sterile land" that is the Caribbean (not the Antilles), but this particular negation expresses in fact the "terrible silence of Man." The "tamtam in the brush" is, Césaire writes, "man speaking" but struck by a "secular burden [le séculaire accablement]." The list of what is missing here (in the Antilles) begins with "no city [ville]." Here is a poet who has left Paris behind, but also carried its image and promise with him. Civilization, in the end, is what is missing, and here Césaire gives us a definition: "this projection of man onto the world; this shaping of the world by man, this striking of the universe into the effigy of man" (Césaire and Ménil 5). The presentation ends, however, not with the heroic re-making of the universe, but with the image of light and encroaching darkness. Much of Césaire's task in *Tropiques* was introductory—to be "un homme d'ensemencement" (76)—and what he introduces is the idea, the image, and the myth, which has neither cause nor effect exclusively in the realm of ideas. This is, perhaps, a way beyond seeing the periphery only as a negation of the centre. This is the work of the poet who is not, like Baudelaire, an outcast from society, or like Rimbaud a meteor exploding across the sky, but rather the anonymous poet whose words are to become, eventually, the law of a new civilization—that is, Vico's poet.

Conclusion: poetry and knowledge

In the 1944 lecture 'Poesie et connaissance', delivered to an international philosophical conference in Port-au-Prince, Haiti, Césaire might be glossing Vichian poetic wisdom.[22] And this text, a manifesto of sorts, is also a demonstration of contradictions internal to Césaire's programme. These contradictions are not best understood as reducible to the particular/universal. Césaire draws from André Breton's Second Manifesto of Surrealism a passage that might be taken to identify the Surrealists—and therefore also Césaire—with the desire for absolute, subjectless knowledge.[23] But then, commenting on the passage, "at the foundation of poetic knowledge . . . all the habits forged, all the images received or seized, all the weight of the body, all the weight of the spirit . . . the body is no longer deaf or blind" (1381–1382). Poetic knowledge is, then, something mystical, an ecstatic unity achieved through total commitment of one's being. Césaire drew literary examples from Proust, Giraudoux, and—although he was cut in later versions of the text—Claudel. This unity is to be achieved by giving up on the search, typical of scientific knowledge, for the useful. "Here, superiority of the animal. And more so of the tree . . . and because the tree is stability, the tree is also surrender. Surrender, to vital movement, to creative energy [élan]" (1383). We recognize

Frobenius, with his schemata of man-plant, man-animal, echoed in the *Cahier* and by Suzanne Césaire in *Tropiques*, and also the Bergsonian slogan of the *élan vital*. The role of the true poet, of true poetry, is to "save humanity" by returning it to its place in the larger natural world. This is, in a phrase that will be repeated in the more famous *Discourse on Colonialism*, an "expansion of man to the measure of the world . . . all great poetry, without ever renouncing humanity, at a very mysterious moment ceases to be strictly human in order to be really cosmic" (1383). This human-beyond-human is brought forth by

> the words [la phrase] of the poet: the primitive words; the universe acted out [joué] and mimed. And because in all true poems, the poet acts out the world, the true poet wants to surrender the word to its free associations, sure that this is in fact to surrender it to the demands of the universe.
>
> (1384)

Césaire seems to have resolved the problem of mere performance that concerned him earlier. He now distinguishes between poetic and scientific knowledge, and links the former to the image.

Even as Césaire formalizes aesthetic and philosophical positions that are most congruent with Vico, he also adopts the most eruptive, individualist, elements of Surrealism. Particularly worth highlighting is the ambiguous nature of the human here as in Vico. The human expands, as it were, by virtue of the divine, to swallow up the natural world. *Physis* and *thesis*, as Auerbach suggested, ceases to be a useful distinction. And yet there is on the face of it nothing in this lecture that has to do with the colonial, nothing that would mark Césaire's subject-position as subaltern. Nothing, one might say, that has anything to do with elaborating, as he had suggested a decade ago the revolutionary philosopher should do, "techniques of liberation."

The cultural politics of Negritude was an effective pluralization of the Marxist framework and also of the French framework. Césaire had to operate both within and against a framework of republican universalism that was structured by racial hierarchy. He sought to counter it with a materialism that was not equally oppressive. Brennan highlights the contemporary relevance of the choice between ostentatiously abandoning any kind of humanism and another path that would attempt the more difficult construction of a new humanism—perhaps one made "to the measure of the world." Césaire's project intersects with Vico's *New Science* in that both are attuned to emotion, to physicality, to the material of collective life, but also to language as a powerful force in shaping—limiting and expanding—human experience. They share a non-rationalist, which is not to say an irrational, universalism.

But there are very important differences. Césaire's poetic knowledge is eruptive, prospective.[24] He is a product of a culture of the avant-garde, and that is not at all Vico's world. Césaire's investment in Mallarmé, in the Surrealists, in the hermeticism of poetry as a tool to explode, or at least reshape, the French language goes hand in hand with his insistence that it is only from within this language, within the sphere of *francophonie*, that concrete freedom is really available for the Antillean. Césaire was, Wilder reminds us, always beholden to an elite-oriented model of politics, and this elitism is equally evident in his literary production (292). From the perspective of the Vichian tradition Brennan has sought to reconstruct, Césaire is thus a useful and challenging figure. His hermetic and avant-gardist investments do not fit well into the demotic aesthetic Brennan valorizes. His example nonetheless suggests that theatricality as a literary technique might be put into contrast with the ironic detachment and rhetorical displacements that Brennan finds essential to, for instance, Nietzsche and Bataille.

Césaire's 1944 lecture is a summation, a programmatic statement. It is a good place to break off our treatment, and for several reasons. Chief among them is that as the war in Europe ends, Césaire will enter intensely into politics. Although the material examined above provides, I hope, a useful background for it, departmentalization itself raises questions and problems that cannot be treated here.[25] Finally, it is after 1945 that, as Arnold and others have noted, Césaire began to work closely with actual Marxists. Thus not only *Discourse on Colonialism,* but also Césaire's later theatrical work, must be considered elsewhere.

Notes

1 All translations, unless otherwise noted, are my own.
2 As far as I know, Banchetti-Robino's paper is the only other attempt to link Césaire and Vico. It will be clear that my approach is significantly different in both its reading of Vico and of Negritude more broadly, especially regarding the usefulness of the term 'historicism' in this context.
3 "'Verum' et 'factum'... convertuntur." First articulated in *De antiquissima*, attributed there to the ancient Italian sages (Berlin and Hardy 35n). The literature on Vico is enormous. I believe Isaiah Berlin's essay is still a useful place to begin, as are the introductory essays to the Bergin-Fisch translations. More recently, and useful here: Bull; Marshall, *Transformation*; Naddeo; Remaud; Robertson. See also, indicating the variety of possible approaches, Lilla; Luft; Mali, *Rehabilitation*; Stone.
4 Berlin glosses this point in the following way, Vico "does not account for our knowledge of other selves—individual or collective, living or dead—by invoking the language of empathy, or analogical reasoning, or intuition, or participating in the unity of the World Spirit. That has been left to his interpreters. He rests his case on his conviction that what men have made, other men, because their minds are those of men, can always, in principle, 'enter into'." And in the note, contra Croce, Gentile, Collingwood: "The influence of Vico's central principle on the German philosopher Wilhelm Dilthey and on the French historian Jules Michelet, and, less directly, on social anthropologists, philologists and historians of culture, has, as a rule, taken more empirical and less speculative forms" (Berlin and Hardy 47, 47n2).
5 And Auerbach goes on: "Vico's poetical age, the golden age, is not an age of natural freedom, but an age of institutions."
6 For some suggestions in this direction, see Berlin and Hardy (43–44). Berlin is most interested here, for his own reasons, in Vichian ways of distinguishing between the natural and the human sciences. However, it is easy to see how this position would allow a Vichian to accept a great many ideas about the biological limits on human being; indeed, to accept much racial science discourse, even if rejecting its central historical meaning. That, in fact, is also what Negritude seems to have done with the discourse of race from the beginning.

7   These is a large claim, but for one recent appraisal, see Goldstein.
8   For very rapid overviews of Vico reception up to the 1940s, see the lucid remarks in the introduction to the Bergin-Fisch translation of the *Autobiography* (Vico et al.). Relevant here is Pons, "Vico in French Thought" in Tagliacozzo and White. See also Marshall, "Current State." On Vico and philology see Leersen, "The Rise of Philology" in Bod et al. Auerbach worked on Vico early in his career; his German translation of *New Science* appeared in 1924. The editor for the project, Gottfried Salomon, would a few years later make the German translation of Sorel's *Reflections on Violence*.
9   Michelet made the first translation into French, not long after one had been made into German. In 1844, the princess Cristina Belgioioso, an altogether remarkable person, made another substantially more exact translation. She wrote of Michelet that his faithfulness prevented him from clarifying anything, but not from changing things (Vico and Belgioioso cxvii–cxviii). It seems that Marx had the Belgioioso translation. See "Marx's Relation to Vico: A Philological Approach" in Tagliacozzo. For Michelet on Vico, see the chapter in Mali, *The Legacy of Vico in Modern Cultural History*. For a broader contextualization of Belgioioso, see Smith.
10 Bachofen is one excellent example of how close it was possible to come in the nineteenth century to Vico without any of the political affinities that interest Brennan (Gossman).
11 There is a standing disagreement about the degree to which Césaire in his early years ought to be seen as a Marxist intellectual. One way of resolving or displacing this disagreement is to say that Césaire was working within a Vichian-Marxist position, one concerned with a certain relationship between materiality and culture, rather than a simply Marxist one. The most important conditioning factor here is the context of French political culture and institutions. Arnold, whose authority over the corpus of Césaire is hard to contest, argues against seeing Césaire, especially in his earlier years, as a Marxist: "from 1935 to 45, the materialist dialectic could not be further from Aimé Césaire's intellectual references" (Césaire 1281). Other scholars want to emphasize the Marxist horizon from the beginning (Miller, "The (Revised) Birth"; Nesbitt).
12 See recently Arnold, "Césaire Is Dead." But also the extended discussion in Arnold, *Modernism and Negritude*.
13 And not just any communism. When Césaire arrived in Paris in 1931, the PCF and the International generally was committed to the 'Third Period' line, which, in view of the imminent and immanent crises of the capitalist world, rejected any cooperation with other elements on the Left. On this moment, I have found useful the essays in Worley. The special function of Paris and London in anticolonial politics as imperial centres of action has received significant attention recently (Boittin; Matera).
14 Scans of the surviving copies of the original publication are available online, but I will refer here to Arnold's critical edition.
15 Arnold gives the source as the 1877 edition, from which I also quote.
16 This text has recently received more attention because it seems to be the first appearance of the term *négritude*. It was, for instance, republished in *le Temps modernes* (Césaire, "Nègreries"). On the Marxist references and contextualization of this text, see Miller, "The (Revised) Birth."
17 There are two texts on Péguy: first in a PCF journal in Martinique, and second in *Tropiques*. More attention has been paid to the second, even while it has been brushed aside as a sop to the censors. Certainly Péguy was a nationalist hero, but he wasn't *only* that. His popularity in the interwar is astounding. Péguy fascinated Walter Benjamin and appears as an essentially French icon of attachment to the *patrie* beyond politics, for instance in Marc Bloch's *l'Étrange défaite*.
18 Arnold, again, suggests: I to 31, II to 64, III to 91, and the remainder IV (73).
19 On this issue, see Miller, "Editing and Editorializing: The New Genetic Cahier of Aimé Césaire."

20 Square brackets refer to the strophe as numbered by Arnold in the critical edition, but also in his recent facing translation.

21 It is hard not to see here an intertext in Wordsworth's sonnet on Toussaint. Wordsworth's daffodils, interestingly, appear in David Scott's interview with Sylvia Wynter as *the* image of an irrelevant or weaponized European culture that Anglophone Caribbean intellectuals were obliged to enter into (Scott and Wynter).

22 On the context and mechanics of this meeting, see the essays—especially François'— collected together with original texts from the conference, in *Moun – Revue de philosophie* IV.7, Janvier 2008.

23 This is one major characteristic of a certain strand of French philosophy, one which, in the interwar period and especially after, mobilized Spinoza to defend rationalism against, especially, Heideggerian phenomenology, with ambiguous consequences, for instance, in the work of Althusser. See Peden.

24 The image is, as he said later, "un dépassement." The Vichian image is something to be read and excavated from its cultural ground by the philologue. On the other hand, one might say that there is a reminder of the *image* that crystallizes the Sorelian myth in Césaire. This would probably be going too far, ignoring the long line of symbolist thinking on the power of the poetic image on which Césaire is manifestly drawing—although the symbolists of the 1880s and 90s were not without contacts with the anarchist milieu out of which, at least partly, Sorel's thinking sprang. See the interview with Leiner at the beginning of the 1978 reprint: Césaire and Ménil (xviii).

25 For the immediate postwar moment, Gary Wilder's *Freedom Time* is certainly the major recent scholarly statement.

Works cited

Arnold, A. James. "Césaire Is Dead: Long Live Césaire! Recuperations and Reparations." *French Politics, Culture & Society*, vol. 27, no. 3, 2009, pp. 9–18.

—. *Modernism and Negritude: The Poetry and Poetics of Aimé Césaire*. Harvard UP, 1981.

Auerbach, Erich. *Scenes from the Drama of European Literature*. University of Minnesota Press, 1984.

Banchetti-Robino, Marina Paola. "Black Orpheus and Aesthetic Historicism: On Vico and Negritude." *Journal of French and Francophone Philosophy – Revue de la philosophie française et de langue française,* vol. 19, no. 2, 2011, pp. 121–135.

Berlin, Isaiah, and Henry Hardy. *Three Critics of the Enlightenment: Vico, Hamann, Herder*. Princeton UP, 2000.

Bod, Rens, et al., editors. *The Making of the Humanities. Volume II: From Early Modern to Modern Disciplines*. Amsterdam UP, 2012.

Boittin, Jennifer Anne. *Colonial Metropolis: The Urban Grounds of Anti-Imperialism and Feminism in Interwar Paris*. University of Nebraska Press, 2010.

Brennan, Timothy. *Borrowed Light: Vico, Hegel, and the Colonies*. Stanford UP, 2014.

Bull, Malcolm. *Inventing Falsehood, Making Truth: Vico and Neapolitan Painting*. Princeton UP, 2013.

Césaire, Aimé. "Nègreries: Conscience raciale et révolution sociale." *Les Temps modernes*, no. 676, 2013, pp. 249–251.

—. *Poésie, théâtre, essais et discours: Édition critique*. Edited by A. James Arnold, CNRS éditions, 2014.

—, and René Ménil. *Tropiques*. Edition Jean-Michel Place, 1978.

Cheymol, Marc, and Philippe Ollé-Laprune, editors. *Aimé Césaire à l'oeuvre: actes du colloque international*. Archives contemporaines, 2010.

Daut, Marlene L. "Before Harlem: The Franco-Haitian Grammar of Transnational African American Writing." *J19: The Journal of Nineteenth-Century Americanists*, vol. 3, no. 2, Fall 2015, pp. 385–392.

Davis, Gregson. *Aimé Césaire*. Cambridge UP, 1997.

Edwards, Brent Hayes. *The Practice of Diaspora: Literature, Translation, and the Rise of Black Internationalism*. Harvard UP, 2003.

Gilroy, Paul. *The Black Atlantic: Modernity and Double Consciousness*. Harvard UP, 1993.

Goebel, Michael. *Anti-Imperial Metropolis: Interwar Paris and the Seeds of Third World Nationalism*. Cambridge UP, 2015.

Goldstein, Jan. "Toward an Empirical History of Moral Thinking: The Case of Racial Theory in Mid-Nineteenth-Century France." *The American Historical Review*, vol. 120, no. 1, Feb. 2015, pp. 1–27.

Gossman, Lionel. *Basel in the Age of Burckhardt: A Study in Unseasonable Ideas*. University of Chicago Press, 2000.

Holt, Thomas C. *The Problem of Freedom: Race, Labor, and Politics in Jamaica and Britain, 1832–1938*. Johns Hopkins UP, 1992.

Larcher, Silyane. *L'autre citoyen. L'idéal républicain et les Antilles après l'esclavage*. Armand Colin Editions, 2014.

Lilla, Mark. *G.B. Vico: The Making of an Anti-Modern*. Harvard UP, 1993.

Luft, Sandra Rudnick. *Vico's Uncanny Humanism: Reading the New Science between Modern and Postmodern*. Cornell UP, 2003.

Mali, Joseph. *The Legacy of Vico in Modern Cultural History: From Jules Michelet to Isaiah Berlin*. Cambridge UP, 2012.

—. *The Rehabilitation of Myth: Vico's New Science*. Cambridge UP, 1992.

Marshall, David L. "The Current State of Vico Scholarship." *Journal of the History of Ideas*, vol. 72, Jan. 2011, pp. 141–160.

—. *Vico and the Transformation of Rhetoric in Early Modern Europe*. Cambridge UP, 2010.

Matera, Marc. *Black London: The Imperial Metropolis and Decolonization in the Twentieth Century*. University of California Press, 2015.

Ménil, René. "Une doctrine réactionnaire: La Négritude." *Action: Revue théorique et politique du Parti Communiste Martiniquais*, no. 1, Aug. 1963, pp. 37–50.

Michelet, Jules. *Le peuple*. Calmann Lévy, 1877.

Miller, Christopher L. "Editing and Editorializing: The New Genetic Cahier of Aimé Césaire." *South Atlantic Quarterly*, vol. 115, no. 3, July 2016, pp. 441–455.

—. "The (Revised) Birth of Negritude: Communist Revolution and 'the Immanent Negro' in 1935." *Pmla*, vol. 125, no. 3, 2010, pp. 743–749.

Naddeo, Barbara Ann. *Vico and Naples: The Urban Origins of Modern Social Theory*. Cornell UP, 2011.

Nesbitt, Nick. "From Louverture to Lenin: Aimé Césaire and Anticolonial Marxism." *Small Axe*, vol. 19, no. 3, Dec. 2015, pp. 129–144.

Peden, Knox. *Spinoza Contra Phenomenology: French Rationalism from Cavaillès to Deleuze*. Stanford UP, 2014.

Remaud, Olivier. *Les Archives de l'humanité: Essai sur la philosophie de Vico*. Seuil, 2004.

Robertson, John. *The Case for the Enlightenment: Scotland and Naples 1680–1760*. Cambridge UP, 2005.

Scott, David, and Sylvia Wynter. "The Re-Enchantment of Humanism: An Interview with Sylvia Wynter." *Small Axe*, vol. 8, Sept. 2000, pp. 119–207.

Smith, Bonnie G. *The Gender of History: Men, Women, and Historical Practice.* Harvard UP, 2000.

Stone, Harold Samuel. *Vico's Cultural History: The Production and Transmission of Ideas in Naples, 1685–1750.* Brill, 1997.

Tagliacozzo, Giorgio ed. *Vico and Marx: Affinities and Contrasts.* Humanities Press, 1983.

—, and Hayden V. White eds. *Giambattista Vico: An International Symposium.* Johns Hopkins Press, 1969.

Vico, Giambattista, et al. *The Autobiography of Giambattista Vico.* Cornell UP, 1944.

—, and Cristina Belgioioso. *La Science nouvelle.* J. Renouard et cie, 1844.

Wilder, Gary. *Freedom Time: Negritude, Decolonization, and the Future of the World.* Duke UP, 2015.

—. *The French Imperial Nation-State: Negritude and Colonial Humanism between the Two World Wars.* University of Chicago Press, 2005.

Worley, Matthew. *In Search of Revolution: International Communist Parties in the Third Period.* I.B. Tauris, 2004.

3 Rabindranath Tagore's postcolonialism

A vision of decolonization and a modernist idealism

Himani Bannerji

Naming is an act of clarification, as it involves a thorough scrutiny of the object to be named. This is important for characterizing the social thought of Rabindranath Tagore, which has been a matter of controversy for almost a century. It has been considered paradoxical because, while a critic of colonialism, he rejected nationalism as a means for decolonization and the emerging national state as an expression of true freedom. Consequently he has been thought of variously as antinationalist, a supporter of British rule, a derivative thinker influenced by the West, a good nationalist who served as the conscience of the nation, a cosmopolitanist, and a humanist philosopher, as well as a mystic and a spiritual guide. This confusion regarding Tagore's social thought could be overcome by considering him a modernist postcolonial thinker, which would allow us to achieve a wider scope within which we could grapple with the complex, at times contradictory, and even unresolved nature of his critique.[1]

The notion of postcolonialism has two aspects: one descriptive and the other epistemological with political consequences. As a descriptive notion it lacks any specific content other than a reference to a historical period of colonialism and steps taken to end it. Even though this notion has been generally associated with postmodernism, there is no particular reason why it cannot instead be modernist. The notion has no predetermined theoretical and methodological direction. It serves as a general term which then takes on a theoretical valence depending on the critic's own knowledge standpoint. The notion of 'post' implies an ambiguity as to whether we are speaking about what follows colonialism politically and socially, or about the politico-ideological methods of decolonization undertaken. Thus postcolonialism could imply both what Frantz Fanon called a "true decolonization" and a "false decolonization" (neocolonialism).[2]

Here we need to provide a context for the emergence of the notion of postcolonialism and of postcolonial studies. The mid to late 1970s (Edward Said's *Orientalism* was first published in 1978) could be taken as the time of their emergence as a type of literary/cultural critique displacing Commonwealth literature.[3] Mainly used by academics, these lacked any direct connection to anticolonial/antiimperialist movements and writings of the time. While they challenged colonial hegemonic discourses of the time, the notions of postcolonialism and postcolonial studies also blocked the anti-colonialist/imperialist critiques created in the course of socialist and communist liberation struggles, especially in the third world. This was at once a radical and a restraining move, resulting in the displacement of a historical materialist or marxist approach by versions of liberal thought. The problematic of colonialism was phrased in discursive terms, without a constitutive relation between colonial discourse and expanding reproduction of capital from Western metropoles, resulting in colonial social formations and cultures. The issue of hegemony, mediated through everyday and official practices, such as ideologies of gender, racialization, and 'othering' practices of difference, was delinked from class. Internal differentiations within the colonized societies and their social repercussions were occluded in favour of decolonization of the mind rather than of the society as a whole. Performing 'decolonization' within this discursive approach amounted to creating a counter discourse of cultural nationalism and relying on re-reading or the invention of pre-colonial cultural forms and traditions.

As an alternative to a marxist, historical materialist approach, this reading of postcolonialism tilted towards postmodernism, a coincidental ideological phenomenon arising in the 1970s in opposition to theoretical marxism.[4] Rejecting both marxism and Enlightenment as variants of modernism, postmodernism became the commonsense understanding of postcolonialism. But this commonsensical association is not dictated by any logic from within the notion of postcolonialism itself. Currently there is a necessity to appropriate the notion to modernism because of relativism that deprived postmodernism of the basis for a genuine critique. But more particularly it is important to assert the value of modernism for a proper assessment of the literature and social thought of the colonies in a pre-marxist time, when marxist ideas and practices had not been generalized. The constellation of ideas and practices coded by the notion of modernism have once more become important in challenging the various right-wing groups seeking national renewal.[5] For Tagore and other early Asian modernists, decolonization enjoined universalist humanism and pedagogical methods understood very broadly with suitable institutions. This modernism simultaneously advocated agricultural and artisanal cooperatives as an alternative to industrial capitalism, along with a universalist culture. Tagore sought decolonization of consciousness and civil society without resorting to nationalism. His antinationalist universalist humanism was an idealist project, but not totally abstract, because he created this worldview with a deep awareness of the socio-political situation in the colonies and in the world elsewhere. His critique of colonialism still has a global relevance.

Tagore's rejection of nationalism, especially through armed struggle and other modes of violence, dates back to approximately 1907. From this time he elaborated the notion of 'constructive *swadeshi*'—in opposition to nationalism—for bringing about a genuine decolonization.[6] He synthesized socio-cultural resources from India, Europe, and elsewhere for his project of a new humanism, which would provide a substantive social being and motives for thought and action without reference to the premises of colonialism. This new subjectivity and agency would not be reactive or reactionary, both of which responses Tagore associated with nationalism. His social thought is comparable to that of Frantz Fanon, who decades later developed his own critique of colonialism in some similar ways. Unlike Fanon, a psychoanalyst and a combatant in the Algerian revolution who 'stretched' Marx in order to understand colonialism's visceral relation to capital, Rabindranath did not have such a socialist/communist position. He became gradually clearer about the capitalist nature of colonialism as he frequently travelled to Europe between 1912 and 1930 and was directly exposed to the First World War, the rise of Bolshevism, the Russian Revolution, and the rise of fascism and imperialism in Italy and Japan and of National Socialism in Germany, until his death in 1941. Through these decades his modernist postcolonialism with its socio-historical awareness began to approximate aspects of a socialist critique.

Speaking of Indian resources, we can connect Tagore's modernism to the project of social reform that arose in nineteenth-century colonial Bengal. This period saw various social and ideological changes advocated by elite sections of Bengali society in relation to religion, 'education', caste, and the treatment of women.[7] The most noteworthy aspect of this social reform is the establishment by Raja Rammohan Roy, Dwarakanath Tagore, and others, of the Brahmo Samaj (1861), centred on a monotheistic philosophical interpretation of hinduism discarding brahminical idolatry and ritualism. Though this movement arose in the context of colonialism, it was not developed under the influence of christianity, but rather propagated universalism and humanism found in the Upanishads,[8] with a syncretic admixture of islamic rationalism.[9] Seen by many as a colonial discourse/'derivative' thought (Chatterjee), the discourse of social reform and the central tenets of Brahmo Samaj provided a basis of reception to reformist values of European Enlightenment. They shared the principle of reason as a primary human capacity. Emerging from the European Renaissance, the Enlightenment was characterized by ideas of reason, humanism, the idea of the individual, scepticism, and belief in science. It made truth claims regarding what reason could discover, and also contained a philosophy of nature. Politically, it was marked by ideas of liberty, equality, and fraternity, which provided the basis for the political thought of the French Revolution. As had happened earlier in Indian history, Bengal's modernist social reform also called for an overall transformation of the civil society and challenged brahminical casteist hinduism by advocating liberal morals and manners involving family, male–female relations, women's education and social participation, and a secular outlook on religious and ethnic differences. This modernism also materialized in demands for legislative changes such as the abolition of the practice of *sati* (burning of women on their husband's funeral pyre), ending child marriage, raising the age of sexual consent, and the right to widow remarriage. This modernist trend was vigorously contested by hindu cultural revivalism, which—contrary to the current postcolonial claim—was not anticolonial, and at times actually opportunistically supported by the colonial state. This practice rested in the colonial governing policy of divide and rule by creating religious/political constituencies, mainly hindu and muslim. Revivalism partially morphed into cultural nationalism, leading to the *swadeshi* anti-partition movement following the partition of Bengal in 1905, which unleashed communal riots and loss of life and property of mainly muslim traders and peasants. For the rest of his life Tagore constantly engaged in debate and dialogue with hindu cultural revivalism and colonial discourse.

Tagore's modernism and anti-nationalism do not have a simple linear relationship. Norms and forms that we call 'modernity' today were developed through his life. Modernity does not have an unequivocal meaning but rather is marked by ambiguities and antinomies (Mazumdar and Kaiwar; Berman; Williams, *Modern Tragedy*). Deriving from European Enlightenment connected to the Renaissance, the term has two aspects: philosophical, consisting of concepts such as universalism, humanism, and the individual in their ideal form; and socio-historical, where they are restricted and modified through actual usage. As the Enlightenment emerged with the decline of feudalism through the rise of capitalism and the development of the bourgeoisie, modernism provided the legitimation apparatus of the rule of the new class in its bid to represent all non-aristocratic social strata and to lead the discontented subaltern classes of feudalism. In practice, modernist universalism served as the ideological cover of the particularist interests of the bourgeoisie. Yet the ideas themselves constellated within 'modernism' were in keeping with the creation of a just society where liberty, equality, and democracy prevailed.

The failure of the universalist dimension of modernity is directly related to capitalism, colonialism, and the rise of colonial and imperialist nation-states. Its secular universalism was subjected to the cause of power, rendering meaningless the notion of the human and also perverting the idea of Reason solely to a techno-industrial capital-oriented rationality. This debased form of thinking was generalized throughout society as common sense—a sensibility—even where it did not actually result in industrialization and the development of local capitalism. For example, the monocultural plantations for the world market, their labour organizations and labour processes, display the most brutal application of the order of the early factory forms. This modernity is capitalist modernization—a process that continues after the formal ending of colonialism. Lacking the larger philosophical dimension and subjugated by social relations of power, it gives rise to forms of particularist, fragmentary, and self-interested consciousness. It brutally restricts the application of the category human.[10] Hegemony of modernization is directly coercive, though at times creating an apparatus of consent through education systems, governing, and the use of the English language. The primary instrument of colonial capitalist modernization is the nation-state. It is on these bases that Tagore rejected the ideology and politics of nationalism in India and elsewhere. He saw their combination as the primary source of imperialism and fascism. In this regard we need to emphasize Rabindranath's letter in 1926 to C.F. Andrews from Vienna (Dutta and Robinson, *Selected Letters*). Here he refutes his earlier pro-Mussolini views upon a better understanding of the dictatorship prevailing in Italy, which he visited:

[T]he method and principles of Fascism concern all humanity, and it is absurd to imagine that I could ever support a movement which ruthlessly suppresses freedom of expression, enforces observances that are against individual conscience, and walks through a bloodstained path of violence and stealthy crime. I have said over and over again that the aggressive spirit of nationalism and imperialism, religiously cultivated by most of the nations of the West, is a menace to the whole world. The demoralisation which it produces in European politics is sure to have disastrous effects, especially upon the peoples of the East who are helpless against the western methods of exploitation.

(333)

He went on to say that Italy was taught its fascist principle by "modern schoolmasters in America. This has suggested to my mind the possibility of the idea of Fascism being actually an infection from across the Atlantic" (334).

In light of these two faces of modernity, we can make sense of both Tagore's admiration for Europe and its condemnation. He saw deformation in its greedy use of machinery, the reduction of science into exploitive technology, and the state as the prime mover of this soul- and imagination-destroying combination. This is why he could speak of the 'good' English and the 'bad' English, the emblematic forms of which we can find respectively in lifelong friends, such as C.F. Andrews, E.J. Thompson, Romain Rolland, and others, and European fascism. He saw American capitalism standing on the plinth of slavery, and its cultural self devoured by money and the market. In this context Rabindranath was massively disappointed with Japan, from which he once sought support for his world vision. He never turned to the political classes of any country, but instead turned for hope to artists, especially poets and universalist philosophers.

Let us look now more closely at Tagore's views on nationalism, which allowed him no space to imagine humanity or sociality. They are most accessible to modern Western readers through his prose works—his translated novels, letters, and essays—and some of his own creative and critical writings in English. Especially useful are his essay *Nationalism* (1916); his correspondence, particularly with Gandhi, C.F. Andrews, Elmhurst, E.J. Thompson, Romain Rolland, and Yone Noguchi; four novels, *Gora* (1907–1909), *Ghare Baire* (Home and the World), *Chaturanga* (Four Quartets) (1914–1917), and *Charadhyay* (Four Chapters) (1931); and three plays, *Achalayatan* (The Immovable Edifice) (1911), *Muktadhara* (Free Waters) (1922), *Rakta Karabi* (Red Oleander) (1925), and the radio play *Dakghar* (The Post Office) (1912).

Rabindranath's vision of decolonization can be grasped from the following passages, the first from 1908, and the second from the 1921 inauguration of his university, Visva Bharati (world university). About Bengali/Indian nationalism he says:

Some of us are reported to be of the opinion that it is mass animosity against the British that will unify India . . . So this anti-British animus, they say must be our chief weapon . . . If that is true, then once the cause of animosity is gone, in other words, when the British leave this country, that artificial bond of unity will snap in a moment. Where then shall we find a second target for animosity? We shall not need to travel far. We shall find it here, in our country, where we shall mangle each other in mutual antagonism, a thirst for each other's blood.

('*Path o Patheyo*' [Ends and Means], qtd. in Dutta and Robinson, *Myriad-Minded Man* 152)

Tagore seems to speak as a prophet, but his use of the future tense is actually born out of experiences of the present, of hindu–muslim riots that shook Bengal after Lord Curzon's 1905 partition and of others that followed, based on the politics of 'two nations' popular with colonial rulers and Indian nationalist parties alike. The second statement provides Tagore's ideal of his new centre of higher learning as "the center of Indian culture":

Let me state it clearly that I have no distrust of any culture because of its foreign character. On the contrary I believe that the shock of outside forces is necessary for maintaining the vitality of our intellect . . . European culture has come to us not only with its knowledge but with its speed. Even when our assimilation is imperfect and aberrations follow, it is rousing our intellectual life from the inertia of formal habits. The contradiction it offers to our traditions makes our consciousness glow.

(qtd. in Dutta and Robinson, *Myriad-Minded Man* 221–222)

This openness to 'foreign', i.e., European, cultures is followed in the same speech by a lucid observation on colonial relations that deform this knowledge both in terms of its content and reception. He goes on to say:

What I object to is the artificial arrangement by which this foreign education tends to occupy all the space of our national mind, and thus kills, or hampers, the great opportunity for the creation of new thought by a new combination of truths.

(222)

This obviously cannot be achieved by what postcolonial critics have called 'mimicry'.[11] Rabindranath calls for a different solution:

It is this which makes me urge that all the elements in our own culture have to be strengthened, not to resist the culture of the West, but to accept it and assimilate it. It must become for us nourishment and not a burden. We must gain mastery over it and not live on sufferance as hewers of texts and drawers of book learning.

(222)

Tagore seeks mutuality and creative relations between cultures, an ethos in which a true humanism can be accomplished at the level of morality and creativity. But was

he able to actualize his vision in his creative and critical opus, his pedagogic philosophy and institutions? Or should we say that the success of his proposal largely
lies in articulating this complex proposal itself?

The themes of women, gender relations, and motherhood serve as lenses
for viewing Tagore's understanding of nationalism. Though motherhood was a
central trope in the narratives of nationalism in India and elsewhere (Koonz),
Rabindranath narrated the nation by introducing other emotional and relational
tropes as well. Only once, in his novel *Gora*, did he concentrate on the notion of
motherhood, where the mother stands for India. But he did not use the concept
in a biological, familial sense (Bagchi, *Interrogating Motherhood*). As feminist
critiques show, this is exceptional anywhere, but especially in South Asia, where
hindu cultural nationalism excelled in this maternal and deified representational
mode.[12] In Rabindranath's novels, freedom and bondage, the main motifs of colonial domination, are often imagined through sexual desire and other male–female
relationships. The notions of the individual and individual choice are also articulated to the narrative and serve as the vehicle and the corollary for different
understandings of freedom as the basis of this choice. Thus the story of nationalism is a story of passion—passion of men and women for each other, passion
for the freedom of the nation—which may lead to death. Personal and political
passions are transparently overlaid. The story of passion, however, is not just
posited, but also critiqued. Passion is contained within the frame of reason, and as
such poses a conflict between acquisitive individualism and humanist universalism. Particularist individual sexual and political desires, choices and freedoms
are thus explored through the paradigm of nature and reason/culture, action and
contemplation, passion and compassion, compulsion and choice. These binary
concepts are further complicated by a bifurcated view of both nature and culture.
In Tagore's novels *Ghare Baire* or *Charadhyay*, for instance, nature is accorded
both a Hobbesian-cum-evolutionist/survivalist aspect as well as a spontaneous,
nurturing, and creative quality. While the former is expressed in nationalism and
its political variants, including greed for wealth and domination of nature, the
latter opens out to humanity in empathy and imagination. 'Freedom' in this view
is tempered with reason. This for Tagore is the characteristic basis for decolonization. This humanist assertion, however, is not achieved without a struggle
between the grasping, narcissistic nature and the fully emancipated human. Given
Rabindranath's own history with nationalism, his own struggle to free himself
from nationalist passions, the presence of the same trajectory seems inevitable.
After all, as early as the 1850s, when nationalism had not yet congealed into clear
political ideologies, Rabindranath and a part of the Tagore family actively participated in shaping the major aspects of Bengali 'nationalist' culture, with influence
in other parts of India.[13] By the 1870s ideological responses ranged from social
reformism and hindu revivalism to liberal and economic nationalism, affecting
each other by osmosis and providing the ground for future politics. For example,
hindu revivalism, politicized into cultural nationalism, remained as a strand of the
Indian National Congress. Gandhi added a generous admixture of hindu culture,
thus colouring bourgeois nationalism's liberal forms.

The problematic of nationalism and decolonization is perhaps best captured in *Ghare Baire*. In this parable of sexual and nationalist passion, an extravagant display of particularisms, the female character Bimala, the emblem of India, the country/nation and its peoples, has ultimately chosen the ethics of universality, has chosen the good—*agape*—over *eros*, constructive social and cultural reconstruction or decolonization over a selfish, suicidal, and murderous passion of nationalism. But this realization in Bimala does not come about until nature, in the shape of physical desire and violent politics, sweeps all before it as a hurricane devastates a landscape, which then awaits regeneration. The narrative indicates that for Rabindranath, also, the road to decolonization is never a straight line. If his poet's psyche is the feminine Bimala, then it is through a wrenching struggle that she gives up the heady and heroic fervour of nationalism to choose the reasonability of the quiet and the relative obscurity of the good. But ultimately it is the last, hardest choice that she has to make, life affirming though not spectacular—seeking reconciliation, not antagonism, between the self and the other. Rabindranath's decolonization, his stance of modernist 'post'-colonization, is not a mere rebuttal to or reversal or transvaluation of the values of colonialism. It stubbornly tries to posit ethical/cultural premises which are unconnected to either colonialism or what he sees as its child, nationalism. The project here is to create the society anew on principles of emancipatory modernism.

Ashis Nandy, Dipesh Chakrabarty, and others of the Subaltern Studies Collective have written in the tradition of conventional postcolonial studies, which is postmodernist. They are in a quandary regarding Rabindranath because they neither accept his modernism nor wholly reject nationalism, since traditionalism or anti-modernism is considered by them as true postcolonialism. They are stuck in their anti-modernist and anti-Enlightenment stance. In large measure their defensiveness comes from staying within a culturally essentializing, spatializing equation of modernity and tradition with the West and East, respectively (Chakrabarty). Their use of the terms 'East', 'West', 'tradition', and 'modernity' is ideological, and as such they allow them to occlude divergent contradictory social relations and historical moments. These terms evolved in different contexts and histories. They range from liberatory notions of rights, egalitarianism, humanism, democracy, and individual self-making to colonial discourses of hierarchy, authoritarianism, racism, domination, and subsumption. It is also noteworthy that some of the emancipatory characteristics, for example humanism, in specific historical and political conjunctures can take on an oppressive tone where the ideal is conflated with the actual. This is how humanism within the context of class, slavery, and colonialism becomes racism, as Sander Gilman has so effectively demonstrated. Thus, instead of humanization of the world's population came racialization, which dehumanized non-Europeans. Among the Europeans themselves class and patriarchy created lesser human beings consisting of the working class and women. But we still have to recognize the qualitatively different forms of modernism mentioned above. There is, after all, a real distinction between utilitarian, techno, or instrumental rationalist modernity at the service of capitalism and imperialism, and a modernity of universalism, which refuses to compromise its idealist and anti-empiricist character. This stance, when integrated with Kantian imperatives of 'enlightenment'—of what an individual might know and live by, unaided by received traditions of conduct (Kant)—introduces into the question of the self, self-making, or discovery a completely different set of considerations than the ones posed by those who see modernity by definition as colonial. Rabindranath's modernist idealist view cannot accommodate the Hegelian master–slave paradigm so crucial to anti-modernist postcolonial critics. In his schema the 'other' is not necessarily in an enslaved and antagonistic relationship to the self. Here sociality of a positive kind trumps the sociality of a sado-masochistic nature as projected by conventional postcolonialists. Modernity as a cluster of concepts is actually sensitive both to relations of oppression and their negation.

It is more important to note that from the eighteenth and nineteenth centuries in Europe and Bengal we hear much about humanist and universalist pedagogies for creating a new social consciousness and subject. In this context Bengal and Europe shared their modernist projects with each other. The modernist idea of 'the individual' spurred the project of social transformation and political thought of Europe. It is only in such an environment that the complex, at times contradictory, project of the French Revolution could be undertaken as the world's first attempt at consciously planning an anti-monarchic, anti-feudal revolution, as Marx and Engels proposed (*German Ideology*; *Communist Manifesto*). These European ventures in ideas and politics in no way contradicted the cultural resources and the moral philosophy of universalism that Rabindranath, the Brahmo Samaj, and other reformers gleaned from Indian philosophy. Indian modernism was the shock absorber when colonialism, violently and suddenly, intruded upon Bengali/Indian society. In this openness to new ideas and forms of living, Rabindranath inherited the legacy of Rammohan Roy, Henry Vivian DeRozio, and Iswarchandra Vidyasagar.[14] Contestation between modernist social reform and hindu or islamic revivalism threw Bengal's upper class/caste elite (*bhadralok*) into fervent productivity along with their encounter with the West. Women, hitherto enclosed in the household, came to schools in large numbers, wrote, and became active in social reform and politics (Kumar). Instead of subordination and paralysis of will attributed to the putative colonial subject by conventional postcolonial thinkers, projects of self-discovery and self-making were everywhere.

Tagore's two educational institutions provide good examples of an alternative decolonizing pedagogy. The first, Santiniketan (the abode of peace), a children's school, was established on the family's estate in 1901, and the second, Sriniketan (abode of grace), an agricultural and handicraft institute, in 1921. Tagore wrote the following about Santiniketan in 1912:

> I have it in mind to make Santiniketan the connecting thread between India and the world. I have found a world center for the study of humanity there. The days of petty nationalism are numbered—let the step towards universal union occur in the fields of Bolpur. I want to make that place somewhere beyond the limits of nation and geography.[15]

Santiniketan's motto is still "where the world finds itself within the same nest." Here Rabindranath tried to practise his transformational pedagogy to create a microcosm of his vision of decolonization. Each participant was to aspire to self-emancipation necessary for a humanist society. The presence of scholars and teachers from outside of India—for example, from Europe, the United States, China, Japan, and Tibet, attracted by Tagore's reputation—gave substance to his universalist ideal. Departments of learning hitherto unknown in India, particularly related to Asia rather than to only Europe or England, were opened in Chinese, Japanese, and Tibetan studies. Some European scholars who fled the hostile world of the Second World War and the Nazi holocaust waited out their time in Santiniketan (Aronson).

Rabindranath's interest in young children and adolescents, similar to Rousseau's and others' Romantic view of childhood, was essential to shaping subjects for his humanist vision. The trope of 'the child' (*shishu*) plays a central role in his literary and critical opus. For him, as among the Europeans, childhood serves as a time for complete receptivity, wonder, and creativity. In symbolic clusters or metaphors in hundreds of poems, songs, and drawings, in short stories, children's rhymes, and textbooks, he celebrated the joy, play, and sorrows of childhood as a quintessential moment of consciousness. Nor did he hide from children, as from adults, the destructive dimensions of colonialism, and the perils of colonial imitation. As a critic of colonialist/racist humanism, he wrote the parable of the fox Hou-Hou in *Shey* (He), in which the fox deserts his own world to be accepted as 'human' (Tagore, *Rabindra Rachanaboli*, vol. 9). To this end, he shaved off his dark fur, revealing the pink and white skin underneath, cut off his tail, tottered on two legs, and spoke haltingly in an alien language. Needless to say, his fate was not enviable. This sad and amusing children's story has an uncanny echo of Tagore's 1921 letter to Edward Thompson about winning the Nobel prize in 1913 as a poet of the English language:

> You know I began to pay court in your language when I was fifty . . . In my translations I timidly avoid all difficulties, which has the effect of making them smooth and thin. I know I am misrepresenting myself as a poet to the western readers. But when I began this career of falsifying my coins I did it in play. Now I am becoming frightened of its enormity and I am willing to make a confession of my misdeeds and withdraw into my original vocation as a mere Bengali poet.
>
> (qtd. in Dutta and Robinson, *Myriad-Minded Man* 264)

The universal—especially as imagined and expressed by a poet and an educator—obviously could not be won by ignoring the concretely local and the personal. In this his creative transcendence was the flight of a bird, which has both a nest and a sky. The line of his song, "I am restless, thirsting for the afar," captures this mentality. Thus his notion of the 'human' is one of a consciousness constantly moving outwards, while also being the denizen of a local world. Tagore's use of the notion 'human' should be understood as a category of desire, of always moving towards

a being, rather than one of an empirical achievement that colonial modernization arrogated only to Europeans. In his view the 'others' of Europe, the 'lesser'
people of the colonies, the slaves, were portrayed as lagging behind or being consigned to an essentialism of savagery. For Tagore, however, the becoming of the
human is always incomplete and unfolding as history moves on. This, he holds,
is a universal truth for all humanity. His modernist humanism does not allow the
form and content to absolutely coincide—as opposed to Hegel, for whom the
ideal and the actual do coincide in his conceptualization of the Prussian state as a
full expression of the Idea. For Tagore the ideal towards which humanity moves
always holds out a receding horizon.

Rabindranath's modernism is idealist—that is, not historical materialist. But
it is still a markedly different type of modernism than that of the equally idealist
colonial or postmodernist variety. It is definitely not colonial discourse. The difference between these two modernisms can be seen by comparing Rabindranath's
ideas on the 'new man' with those of Nietzsche. Rabindranath articulated his
concept within a universalist humanist perspective, while Nietzsche's over-man
(*Übermensch*) was shaped within the ethos of capitalist colonialism, class, and
slavery. Nietzsche's 'man' is an atomized egotistic entity who lacks any positive
connection with the rest of humanity. The *Übermensch* 'surpasses' only to the
extent that he negates the notion of sociality. Though the idea of 'going beyond'
is common to both, in Nietzsche it has a touch of social Darwinism that makes the
Übermensch vulnerable to an empiricist twist later by Nazis and other fascists.[16]
This isolationist individualism is the content of the loneliness of Zarathustra, who
despises the sociality of the 'herd'. Similarly, a colonial modernist understanding
of 'civilization' implies that Europe has already achieved it, while the colonized
societies were either decaying or were yet to approach this state. This is the meaning of colonial discourse, which is a combination of racialized anthropology
based on social Darwinism. Evolution thus understood serves as an ideology of
legitimation for capitalist colonization. This is a compendium of colonial power/
knowledge (Gould; Rose). Tagore's idealist modernism, on the contrary, does not
compromise with empiricism. By virtue of that, it has the potential for providing
a critique of the empirically existing here and now. Socially speaking, the 'self'
and the 'other' are not in a dominating relation, but one of aspiring identification. The myth of the over-man is replaced by 'Man' as everyman, and ideas
such as 'purity of race' and ethnicity and 'race heroes' are unutterable within this
framework. Thus, Tagore rejected nationalism, with its elements of ethnicity and
civilizational/cultural supremacy.

Tagore's idea of freedom is one of a socially individualized self. It calls for
a metaphor. He compares this self to a river, which finds its way to the ocean of
humanity. The metaphor is dramatized in Tagore's plays *Muktadhara* (*Rabindra
Rachanaboli*, vol. 5) and *Achalayatan* (*Rabindra Rachanaboli*, vol. 5). The first
play centres on a large dam project, with a fetishistic attitude towards technology
for the domination of nature and inspired by a mechanical rationality. The common people, moved by their spontaneity and sociality, who are led by a visionary
poet, unite against the machine-worshipping bureaucrats to stop their project of

enslaving nature, and their daily and creative life succeed in freeing the river. Nature here is not a primitive force, but a life-giving one to be cooperated with. It is the colonizing of it that turns the river destructive, as life is deformed by processes of domination. Tagore makes a contrast here between the 'order' imposed by a wealth-driven modernization and the 'rhythm' of life and creativity invoking dance. In *Achalayatan*, this same message is played out with regard to students in a school which represses their creativity, curiosity, and joy in learning. This repression is also considered by Tagore as anti-nature. Again the pupils and the poet break down the walls of this sterile institution of dead learning. Once more 'freedom' is conjoined with imagination and sociality, and a concept of learning congenial to children's nature emerges. This learning is connected with an inner need rather than an institutional one. In 1911 in *Jibansmriti* (My Reminiscences) Rabindranath expressed a similar sentiment:

> Only in a land where an animus of divisiveness reigns supreme, and innumerable petty barriers separate one from another, must this longing to express a larger life in one's own remain unsatisfied. I strained to reach humanity in my youth, as in my childhood I yearned for the outside world from within the chalk circle drawn around me by the servants . . . And yet, if we cannot get in touch with it, if no breeze can blow from it, no current flow out of it, no path be open to the free passage of travelers—then the dead things accumulating around us will never be removed, but continue to mount up until they smother all vestige of life.
>
> (qtd. in Dutta and Robinson, *Myriad-Minded Man*, 431–432)

The play *Raktakarabi* (*Rabindra Rachanaboli*, vol. 6) depicts a conflict between humanist modernity and greed-centred modernization, between the human and the machine. The action takes place in a coal mine, at the heart of which subterranean realm sits a king imprisoned by a technology for exploitation of the earth by means of human labour. Here workers become mere numbers and their overseers a part of the machinery of extraction. In this space imagination, life, love, and sociality are as absent as fresh air, greenery, and sky. Here, where the earth is merely its own skeleton, composed of rocks and minerals, enter imagination, spontaneity, and sexual and social desire. They are expressed through a poet/singer, a young woman who is the principle of love and freedom, and a boy, the sign of innocence. Together they embody freedom, beauty, and imagination, and reject a machine-controlled civilization. They shift the movement of the play from darkness towards light, from the interior to the world outside, from a calculus of greed to poetry. They precipitate a human turmoil of feelings and imagination into the ordered world of exploitation. The freedom comes from the catastrophe when the mine is flooded, as the workers rise up against the system. The king and the workers emerge from the dark chambers to join the humanity in a celebration of life. The message is that nature, people, and society are perverted into 'primitivity/savagery' by domination, mechanization, and a culture of rigidity, narrowness, and acquisitiveness. Tagore's approach is reminiscent of Rousseau and Romanticism; but unlike Tagore, Rousseau has a unidimensional view of culture that is negative. Nature is equally so, except that it is all positive. It is 'the chains' of culture that shackle humanity and destroy the impulse for the good and the social (Rousseau; Coletti). Rabindranath's humanist reading of culture redeems it from this solely negative connotation. Here imagination and empathy as capacity for apprehending the universal are the overdetermining faculties binding nature to culture.

To rescue Rabindranath from a vulgar romanticism we need to know that he was not anti-science or technology when they were meant to serve humanity. He counted among his friends scientists, such as Albert Einstein and Jagadish Chandra Bose, and created a school for agricultural science and technology. As his correspondence with Gandhi reveals, they thoroughly disagreed on these matters (Bhattacharya). A positive and non-instrumental view of science is found in Rabindranath's opinion on Gandhi's advocacy of the spinning wheel (*charka*) as a challenge to textile mills. He says:

> [We] fail to rise above the ideology of the *charka*. The *charka* does not require anyone to think: one simply turns the wheel of the antiquated invention, using the minimum of judgement and stamina. In a more industrious vital country then ours such a proposition would stand no chance of acceptance.
>
> (Dutta and Robinson, *Selected Letters* 365)

As early as 1907–1909 in the novel *Gora*, Tagore commented on the ugly reality of obscurantism and primitive agricultural implements oppressing rural India, and the need for bringing to the countryside a scientific and humanist outlook, especially against caste and superstition:

> The ties of the samaj, the devotion to customs, do not give them any strength in practice . . . Gora realized that this samaj gives no help in times of need, no support in the face of danger, it can only harass people by enforcing a rigid conformity . . . In the immobility of the rural life Gora saw the real weakness of our country in an absolutely unadorned form.
>
> (qtd. in Sarkar, *Swadeshi Movement* 62)

Last but not least, Rabindranath wrote a popular science textbook for his students, entitled *Visva Parichay* (Knowing the Universe) (*Rabindra Rachanaboli*, vol. 15).

Rabindranath's modernism did not endear him to the Indian, Japanese, or European nationalists. He himself did not spare them in his furious condemnation of imperialism arising out of nationalism. A great example is to be found in his letter to the imperial Japanese poet, Yone Noguchi, written in 1938:

> When you speak, therefore, of 'the inevitable means, terrible it is though, for establishing a new great world in the Asiatic continent'—signifying, I suppose, the bombing of Chinese women and children and the desecration of ancient temples and universities as a means of saving China for Asia—you

are ascribing to humanity a way of life which is not even inevitable among the animals and would certainly not apply to the East, in spite of her occasional aberrations. You are building your conception of Asia which would be raised on a tower of skulls.

(qtd. in Dutta and Robinson, *Selected Letters* 497)

Indian nationalists considered Tagore's unorthodox depiction of peasants and women in creative and critical writings as a slur on hindu womanhood and social order. His universalism and secular humanism were interpreted as collusion with colonial power. Though he rejected his knighthood after the 1919 massacre of Jalianwalabagh perpetrated by the colonial rulers, hindu nationalists in particular referred to him as 'Sir' Rabindranath. They lacked the insight of Gandhi, who called him the 'Great Sentinel', the conscience of the nation. He was sustained by his stubborn belief in the possibilities inherent in human capacities and had a vision to convey for which he made many trips across the world to share. It is against this humanist vision that he judged the hubris of both colonial modernity and imperialist nationalism. In the gathering darkness of the Second World War, and his own fast-approaching death (1941), he expressed his pessimism in *Shabhyatar Shankat* ("The Crisis in Civilization") (*Rabindra Rachanaboli*, vol. 13). In the two decades before he died he came to see, perhaps moved by Gandhi's moral suasion and the Indian National Congress's relatively non-violent mass mobilization, that there might be a necessity of a formal independence. A few lines from "The Crisis in Civilization" are worth quoting here. They are reminiscent of Romain Rolland's saying about "pessimism of the intellect and optimism of the heart," which Antonio Gramsci adopted as his own motto during his years in Mussolini's prison (*Prison Notebooks*). Rabindranath wrote:

The wheels of fate will some day compel the British to give up their Indian Empire. But what kind of India will they leave behind, what stark misery? When the stream of their two centuries' administration runs dry at last, what a waste of mud and filth will they leave behind them! I had at one time believed that the springs of civilization would issue out of the heart of Europe. But today, when I am about to quit the world, that faith has deserted me.

(qtd. in Dutta and Robinson, *Myriad-Minded Man* 364)

But he also wrote in principled optimism:

And yet I shall not commit the grievous sin of losing faith in man. I would rather look forward to the opening of a new chapter in his history after the cataclysm is over and the air rendered clean with the spirit of service and sacrifice.

(364)

Is there something still for us to learn from Rabindranath's vision of postcoloniality? The answer can only be in the affirmative. Rabindranath both dreamed and acted practically—creating institutions, making critical interventions, and taking advisory roles to Gandhi and the Indian National Congress as well as to European anti-fascist movements. His humanist vision notwithstanding, he learned that one could not fly to decolonization, but had to walk through the mud and grime of actual history. Though he was not a historical materialist, his active intelligence, social awareness, moral responsiveness, and profound imagination constantly grappled with existing enigmas and nightmares of history. The beauty of his vision remains compelling, as does his constant caution against particularism/ narrowness, national chauvinism, and their horrific consequences. Though his grand view of humanism rested in idealism, he also made us feel that the future begins now, that this is worth striving for. His ideas themselves, consonant with those of the Enlightenment, aimed at a true decolonization and remain the right goals, though in an idealist framework. No communist/socialist revolution could be imagined without these core values of humanist universalism. Especially now, in our present world, where ultra-nationalist colonialism in various disguises has emerged with a vengeance—in invasions, occupations, genocides, and the daily death of humanity—in the name of democracy, freedom, god, and ethnic belonging, Rabindranath's voice is as important to us as it was to generations before. The last radio broadcast in Paris in 1940, on the eve of the Nazi entry into the city, was Rabindranath's play *Dakghar* (The Post Office) (*Rabindra Rachanaboli*, vol. 5) in André Gide's translation (Dutta and Robinson, *Myriad-Minded Man* 351). It is about a dying child who awaits the king's messenger who will bring him a letter. Neither he nor we know what will be in that letter—a message of death or of hope? But we still need to wait, because in that act of waiting, which is not simply passive, we affirm our fight against violence and despair.

We return here to the concept of postcolonialism, which is no more than a simple descriptive category that signals to a historical period, with no other content. It is its epistemological affiliation that gives it content. Furthermore, the notion of postcolonialism has in itself no logical affiliation to either modernism or postmodernism. Both of these could co-exist with the social and productive relations of capitalism and imperialism. As such, the projected decolonization could only be partial and mainly at the level of culture and consciousness. Colonialism understood as cultural hegemony could only take us so far and poses the cultural politics of nationalism against equally cultural perceptions of colonialism. But the question that haunts us is whether under the auspices of postmodernism we could conceive of a true decolonization, of any holistic project of social transformation. To answer this question we need to recognize that idealist modernism, unlike the modernism of historical materialism, still considers ideas as the moving power of history. But, unlike postmodernism, this does not prevent it from making truth claims for humanism, the desirability for a socially incapacitated creative individual, justice, and freedom. Its humanism implies egalitarianism because the human is a metaphor of commonality. These ideas are non-negotiable in any project of transformation. As a hyper-subjectivist epistemology, idealist modernism cannot, of course, provide a constitutive relationship between consciousness and the social, and accords them qualitatively distinct realms. Nonetheless it can make claims of truth for the importance of imagination, for grand narratives of profound change of con-sciousness. Postmodernism, in contrast, posits an epistemology, if we can call it that, of indecision. It challenges any truth claim for ideas to be effective for a change on a large scale. It cannot sustain a grand narrative of conscious mak-ing of history for humanity. It denies both subjective and objective truth claims any notion of real knowledge regarding the self and the world and their creative relations. As such, its worldview is fragmentary, relativist, and at best capable of bringing about change in local and particular terms. Within the postmodernist framework, lacking space for history, colonialism takes on an episodic charac-ter, and resistance to it is manifested in terms of subversion, displacement, and opposition of identities through cultural manoeuvres. Here the universal human is replaced by the particular individual without a sense of an irreducible interior self or sociality. With its narrow, barely possible formulation of anticolonial-ism, postcolonialism gets into 'pitfalls of nationalism'—even when professing scepticism. Freedom, liberty, justice, equality, etc. become the elements of an illusory grand narrative, and resistance is an issue-based piecemeal matter. The refusal of anything but a factual truth entrenched within this framework makes not only historical change but even critique untenable. It can only 'imagine' history or episodes of debates with an ineffable 'power'; its violences are only epistemic. But unbeknownst to itself, the postmodernist postcolonial position does uphold a grand narrative, that of colonialism, its own point of birth and departure. Though it cannot explain the phenomenon of colonialism, it relies on that very notion to structure its own premises. Its end, therefore, cannot be going beyond colonialism, because its own epistemological validity lies in the colonial project. Unlike idealist modernism, it cannot provide an impetus for a larger vision, such as that which inspired the French Revolution. As the concept of the social cannot be accommodated by postmodernism, it can only speak in terms of atomistic individuals or, at best, in terms of fragments, each a bounded social/cultural group. If we contrast modernist idealism to postmodernist ide-alism, it becomes evident that the abstract generality of universal ideas of the former is matched by the particularist ideas of the latter. The general identity of the human understood as a set of shared creative capacities, including that of empathy, is replaced in postmodernism by segments of cultural identities. The critical horizon of postmodernism is so narrow that it can only offer simple binaries to the existing colonial ones, thus continuing the project of colonialism in reverse. Postmodernism makes decolonization an inconceivable task.

Tagore's idealist project of decolonization provides direction to a path to true decolonization. His humanism is more than a mere aesthetic trope of desire, because it partially emerges from his awareness of the actual social deforma-tion, economic deprivation, and cultural degradation caused by colonialism and imperialism in India and elsewhere. His idealism made him think redemptively. Human redemption and decolonization became one in his thought. But as for all idealism, the path between the fallen present and the utopian future could not be socio-historically charted. He, therefore, resorts to transcendentalism. Lacking a historical materialist understanding of the concrete process of the making of history, and rejecting any role of politics in human liberation, Tagore could not conceive of how people might bring their consciousness to bear on the actual relations of power facing them. But from his modernist idealism the leap can be made to a struggle for a socialist human transformation. In fact, such a transformation without his humanist vision would not lead to decolonization.

Notes

1   This characterization could apply to other late nineteenth and early twentieth-century Asian critics of colonialism, such as Zhou Shuren (Lu Xun) of China (1881–1936) and Count Okakura Kazuko of Japan (1862–1913). See also the ideas of the 'East' and the 'West' with a special reference to the art of Japan, published on the eve of the Russo-Japanese War. Pan-Asianism, started as a political philosophy against imperialism and colonialism, changed into Japanese imperialism in the late nineteenth and first three decades of the twentieth century.

2   Fanon speaks to this in his chapter on 'The Pitfalls of National Consciousness'. See also Sekyi-Otu, who establishes Fanon as a humanist marxist revolutionary.

3   Ania Loomba in *Colonialism/Postcolonialism* rightly points out that postcolonialism entails two dimensions, one of temporality (history), and the other ideological (politics) (7).

4   See Lyotard, *The Postmodern Condition*, the primary text for postmodernism.

5   On the usefulness of left liberal, secular modernist thinkers, see, for example, Said, *The Question of Palestine*; *Covering Islam*; *Peace and its Discontents*.

6   *Swadeshi* means 'of one's own country', which got loosely translated as 'nationalism'. Rabindranath's own approach to *swadeshi*, dubbed by S. Sarkar as 'constructive', is a moderate political tradition, or what he calls "a moderate and broad" political movement which, "quieter and sometimes non-political in its tone, emphasized patient efforts at self-development, ignoring foreign rule rather than lauding an immediate attack on it" (*Swadeshi Movement* 48). This approach of Rabindranath's is a contrast to "political extremism proper [which] tried to turn the boycott into a campaign of full-scale" political struggle, which "set its sights on immediate independence rather than partial reforms or slow self-generation" (48).

7   See S.C. Sarkar; also S. Sarkar, *Swadeshi Movement*; *Critique of Colonial India*; and Raychaudhuri. The modernist values of the Bengali elite are comparable to the equally elite philosophy of the European Renaissance and Enlightenment.

8   Ancient Indian philosophical texts are also known as Vedantas. They mark a transition from Vedic ritualism to spiritual ideas.

9   From the Mu'tazil tradition, a school of Islamic theology, which denies the status of the Qur'an as uncreated, evolved between the eighth and tenth centuries.

10 See Williams' entry in *Keywords*; also see the entries on civilization, culture, modern, and tradition.

11 In his novel *A House for Mr. Biswas*, V.S. Naipaul elaborates the role of the co-opted colonial middle class in maintaining colonial hegemony. See also Soyinka; Homi Bhabha on 'hybridity'. Tagore's attitude is also not cosmopolitan (Collins), not aggregative, but synthetic.

12 There is a vast literature on the topics of women and nationalism and gender and nationalism. See, for example, Bannerji et al.; Yuval-Davis; Bagchi, "Socializing the Girl Child"; and Jayawardena.

13 For Western readers the best source of information on this is Dutta and Robinson's biography of Tagore, *A Myriad-Minded Man*.

14 Rammohan Roy, DeRozio, and Vidyasagar were noted social reformers of nineteenth-century Bengal. For more details see Sarkar, *Critique of Colonial India*, and Joshi.

15 Qtd. in Dutta and Robinson, *Myriad-Minded Man* 205. For more on the foundation of Santiniketan and Sriniketan see chapters 20, 21. See also Dutta and Robinson, *Selected Letters*, especially the letter to his son, Rabindranath Tagore, 11 October 1917.
16 Some aspects of this have been discussed in David McNally's *Bodies of Meanings*. See the introduction and ch. 1. Also, on Nietzsche's connection with Darwin, see Kaufman. On the project of self-making and the *Übermensch* see Norris.

Works cited

Aronson, Alex. *Brief Chronicles of the Time: Personal Reflections of My Stay in Bengal (1937–46)*. Calcutta, Publisher unknown, 1991.

Bagchi, Jasodhara. "*Anandamath* and 'The Home and the World': Positivism Reconfigured." *Rabindranath Tagore's 'Home and the World': A Critical Companion*, edited by Prabir Kumar Dutta, Permanent Black, 2003, pp. 174–186.

—. Interrogating Motherhood. Sage, 2017.

—. "Socializing the Girl Child in Colonial Bengal." *Economic and Political Weekly*, 9 Oct. 1993, pp. 2214–2220.

Bannerji, Himani, Shahrzad Mojab and Judith Whitehead. *Of Property and Propriety: The Role of Gender and Class in Imperialism and Nationalism*. University of Toronto Press, 2001.

Berman, Marshall. *All That Is Solid Melts into Air: The Experience of Modernity*. Penguin, 1988.

Bhabha, Homi. *The Location of Culture*. Routledge, 1994.

Bhattacharya, Malini. "Gora and the Home and the World: The Long Quest for Modernity." *Rabindranath Tagore's 'Home and the World': A Critical Companion*, edited by Prabir Kumar Dutta, Permanent Black, 2003, pp. 127–142.

Bhattacharya, Sabyasachi, editor. *The Mahatma and the Poet: Letters and Debates Between Gandhi and Tagore 1915–1941*. National Book Trust, 1997.

Chakrabarty, Dipesh. *Habitations of Modernity*. Orient Blackswan, 2004.

—. *Provincializing Europe*. Princeton UP, 2000.

Chatterjee, Partha. *Nationalist Thought and the Colonial World: A Derivative Discourse?* Zed Books, 1986.

Chattopadhyay, Kunal, editor. *Genocidal Pogrom in Gujarat: Anatomy of Indian Fascism*. An Inquilabi Communist Sangathan Publication, 2002.

Coletti, Lucio. *From Rousseau to Lenin: Studies in Ideology and Society*. New Left Books, 1972.

Collins, Michael. *Empire, Nationalism and the Postcolonial World*. Routledge, 2011.

Communalism Combat: Genocide. Year 8, no. 76, Sabrang Communications and Publishing, 2002.

Dutta, Krishna and Andrew Robinson. *Rabindranath Tagore: The Myriad-Minded Man*. Bloomsbury, 1995.

—, editors. *Selected Letters of Rabindranath Tagore*. Cambridge UP, 1997.

Dutta, Prabir Kumar, editor. *Rabindranath Tagore's Home and the World: A Critical Companion*. Permanent Black, 2003.

Fanon, Frantz. *The Wretched of the Earth*. Grove Press, 1968.

Gilman, Sander. *Difference and Pathology: Stereotypes of Sexuality, Race and Madness*. Cornell UP, 1985.

Gould, Stephen Jay. *The Mismeasure of Man*. Norton, 1996.

Gramsci, Antonio. *Selections from the Prison Notebooks.* Edited and translated by Quintin Hoare and Geoffrey Nowell. International Publishers, 1971.

Jayawardena, Kumari. *Women and Nationalism in the Third World.* Zed Books, 1986.

Joshi, V.C., editor. *Rammohun Roy and the Process of Modernization in India.* Vikas Publishing House, 1975.

Kant, Emmanuel. "What is Enlightenment?". *Kant: A Collection of Critical Essays,* edited by Robert P. Wolff. Anchor, 1967.

Kaufmann, Walter. *Nietzsche: Philosopher, Psychologist, Antichrist.* Princeton UP, 1974.

Koonz, Claudia. *Mothers in the Fatherland.* Routledge, 2013.

Kumar, Radha. *The History of Doing: Movements for Women's Rights and Feminism, 1800–1990.* Kali for Women, 1993.

Loomba, Ania. *Colonialism/Postcolonialism.* Routledge, 2005.

Lyotard, Jean Francois. *The Postmodern Condition: A Report on Knowledge.* Manchester UP, 1984.

Marx, Karl and Frederick Engels. *The Communist Manifesto.* Penguin Books, 1985.

—. *The German Ideology.* Progress Publishers, 1968.

Mazumdar, Sucheta and Vasant Kaiwar, editors. *Antinomies of Modernity: Essay on Race, Orient, Nation.* Tulika, 2003.

McNally, David. *Bodies of Meaning: Studies on Language, Labour and Liberation.* State University of New York Press, 2001.

Naipaul, V.S. *A House for Mr. Biswas.* André Deutsch, 1961.

—. *The Mimic Men.* Penguin Books, 1969.

Nandy, Ashis. *The Illegitimacy of Nationalism: Rabindranath Tagore and the Politics of the Self.* Oxford UP, 1994.

Norris, Christopher. *The Truth about Postmodernism.* Blackwell, 1993.

Raychaudhuri, Tapan. *Europe Reconsidered: Perceptions of the West in Nineteenth Century Bengal.* Oxford UP, 1988.

Rose, Stephen, Leon Kamin, and R.C. Lewontin. *Not in our Genes: Biology, Ideology and Human Nature.* Pantheon Books, 1984.

Rousseau, Jean Jacques. *The Social Contract, and Discourses.* Translated by G.D.H. Cole. Everyman's Library, 1968.

Said, Edward. *Covering Islam.* Random House, 1981.

—. *Peace and its Discontents: Essays on Palestine in the Middle East Peace Process.* Vintage, 1995.

—. *The Question of Palestine: A Political Essay.* Times Books, 1979.

Sarkar, Sumit. *A Critique of Colonial India.* Papyrus, 1985.

—. *The Swadeshi Movement in Bengal: 1903–1908.* People's Publishing House, 1994.

—. "*Ghare Baire* in its Times." *Rabindranath Tagore's 'Home and the World': A Critical Companion,* edited by Prabir Kumar Dutta. Permanent Black, 2003.

Sarkar, Susobhan Chandra. *Bengal Renaissance and Other Essays.* People's Publishing House, 1970.

Sekyi-Otu, Atu. *Fanon's Dialectic of Experience.* Harvard UP, 1996.

Soyinka, Wole. *The Interpreters.* Africana Publishing, 1972.

Tagore, Rabindranath. *Rabindra Rachanaboli.* West Bengal Education Department, 1980–1996.

—. Achalayatan. *Rabindra Rachanaboli,* vol. 5, 1922.

—. Charadhyay. *Rabindra Rachanaboli,* vol. 8, 1934.

—. Chaturanga. *Rabindra Rachanaboli,* vol. 8, 1916.

—. Dakghar. *Rabindra Rachanaboli,* vol. 5, 1912.

—. Ghare Baire. *Rabindra Rachanaboli*, vol. 8, 1916.

—. Gora. *Rabindra Rachanaboli*, vol. 7, 1910.

—. Jibansmriti. *Rabindra Rachanaboli*, vol. 11, 1912.

—. Muktadhara. *Rabindra Rachanaboli*, vol. 5, 1922.

—. Shabhyatar Shankat. *Rabindra Rachanaboli*, vol. 13, 1990.

—. Shey. *Rabindra Rachanaboli*, vol. 9, 1937.

—. Raktakarabi. *Rabindra Rachanaboli*, vol. 6, 1926.

—. Visva Parichay. *Rabindra Rachanaboli*, vol. 15, 1937.

—. *Letters from Russia*. Translated by Sasadhar Sinha,Visva Bharati, 1960.

—. *Nationalism: The English Writings of Rabindranath Tagore*, edited by Sisir Das. Sahitya Akademi, 1996.

Williams, Raymond. *Modern Tragedy*. Penguin, 1992.

—. *Keywords: A Vocabulary of Culture and Society*. Croomhelm, 1976.

Yuval-Davis, Nira. *Gender and Nation*. Sage, 1997.

4 Voyages of the self

Muslims as anticolonial subjects in Muhammad Iqbal's philosophy of history

Asher Ghaffar

During the early twentieth century, British idealists travelled from Britain to Germany to study Hegel's philosophy. At the advice of Thomas Arnold, Muhammad Iqbal (1877–1938) followed a similar path, travelling from imperial Britain to Germany to complete his doctorate. In Iqbal's dissertation, *The Development of Metaphysics in Persia* (*Development*), the renowned Indo-Pakistani Muslim poet, philosopher, and political theorist developed a global theory of travelling ideas that echoed Goethe's formulation of world literature:

> No idea can seize a people's soul unless, in some sense, it is the people's own. External influences may wake it up from its deep unconscious slumber; but they cannot, so to speak, create it out of nothing. The full significance of a phenomena in the intellectual evolution of a people can only be comprehended in the light of those pre-existing intellectual, political, and social conditions which alone makes its existence inevitable.
>
> (76)

Goethe's idea of world literature was probably important for Iqbal, as it entailed a reciprocal recognition of other national literatures to enrich one's own national literary tradition, and precluded the rote imitation of the foreign. Like British idealists such as J.M.E. McTaggart (1866–1925)—whom Iqbal studied with and frequently cited throughout his life—the poet-philosopher felt a profound affinity with German philosophy, as he put it: "Germany was a kind of second home to my spirit. I learnt much and thought much in that country. The home of Goethe has found a permanent place in my soul" (qtd. in Durrani 462).

This chapter examines these understudied dimensions in Iqbal's thought and situates his Hegelian 'beginnings' after the Second Boer War.[1] Iqbal is read alongside Continental philosophers who were responding to the same international currents, notably the rise of National Socialism. The poet-philosopher drew on Hegel's ideas in his social and political philosophy as an often unstated, conceptual resource to articulate Muslims as anticolonial subjects in his philosophy of history; and his worldly use of Hegelian ideas can be contrasted with his critique of Nietzsche who sought to displace key Hegelian concepts (Brennan *Borrowed Light*).[2]

Building on Timothy Brennan's ground-breaking reading of Hegel, this chapter recuperates forgotten Vichian elements such as orature and polemic that allow for a more nuanced reading of Hegel and anticolonial intellectual history. In light of this relationship with Hegelian thought, Iqbal can be seen as a pioneering anticolonial philosopher of the Muslim self who—like other major anticolonial figures— borrowed from Western thought to create a revised humanism.[3]

His Vichian anticolonial humanism serves to underscore his own philosophical 'beginnings'.[4] Similar to other anticolonial theorists, Iqbal searched for a renewed non-racial humanism and foregrounded the dialectic between religion and oral culture, which places him firmly in the anticolonial humanism that Brennan limns. The first and second sections of this chapter outline Iqbal's affiliation with and critique of British idealists by situating the poet-philosopher in the debates that informed British idealist philosophers at the turn of the nineteenth and twentieth centuries—chiefly the Second Boer War. The third section closely examines Iqbal's critique of Nietzsche, particularly in relation to the Russian Revolution. The final section suggests that Iqbal's engagement with Bergson allowed him to develop a philosophy of history that recuperates a Hegelian-Vichian concept of time that resonates with philosophers such as Theodor Adorno.

Hegel, British idealism, and Iqbal's concept of self

Although Iqbal is often seen as an Islamist,[5] he can be distinguished from Mawdudi, and certainly Qutb, who rejected Westernization. His encouraging view of the Young Turk Revolution provides a sharp contrast to these views: "The truth is that among the Muslim nations, Turkey alone has shaken off its dog-matic slumber, and attained to self-consciousness" (*Reconstruction* 163). Iqbal understood secularism to be a moment in the development towards a Muslim humanist ideal: "The spirit finds its opportunities in the natural, the material, and the secular. All that is secular is, therefore, sacred in the roots of its being" (155). This also distinguishes him from other Muslim modernists who drew on counter-Enlightenment thinkers such as Heidegger and Nietzsche to emphasize modern malaise and anti-Western sentiment brought about by the loss of cultural roots (Mirsepassi).[6]

In addition to emphasizing Islamic sources, it is also important to under-stand Iqbal's religiosity in relationship to British idealists who were rethinking Christianity in light of Darwinism and the utilitarianism of staunch imperial-ists such as Mill and Bentham. British Hegelian emanation theory attempted to challenge Darwinism by suggesting that the self spiritually evolves: "British idealists developed a distinct form of evolution based on Hegel's notion of emanation. We may call this 'Spiritual' evolution, and it is distinct from its contemporaneous competitors, naturalistic and ethical evolution" (Boucher and Vincent 26). Furthermore, British idealists such as the Hegelian T.H. Green (1886–1882) developed a burgeoning liberal social philosophy that emphasized the immanence of the absolute in the social world to address class disparity in imperial Britain. Such global influences—coupled with Iqbal's experience of racism in Britain—point to the importance of the poet-philosopher's voyages to Britain and Germany.

Iqbal's worldly humanism is particularly evident in his sprawling English work of political philosophy, *The Reconstruction of Religious Thought in Islam* (*Reconstruction*), a wide-ranging work of stunning breadth and erudition that draws from Urdu, Persian, German, French, English, and other sources. The *Reconstruction* was composed as a series of seven lectures written in English that were delivered in Madras, Hyderabad, and Aligarh. This work was published as a book in 1930—the same year that Iqbal delivered his famous Allahabad Address—which outlined the idea of a new Muslim nation, while simultaneously critiquing nationalism. Nonetheless, the Allahabad Address was instrumental in the formation of a new state for Muslims—Pakistan—whose evening ritualistic flag-raising ceremony at its borders would surely have been Iqbal's anathema. *Reconstruction* is a more complicated and nuanced version of Iqbal's political address in which he reads Western thought alongside Islamic philosophy to reconstruct the Muslim subject as open to the vicissitudes of history.

Muslim thought did not vacate Western philosophy for Iqbal, but drew on it to move Islam forward—while also critiquing it from within. The montage-like form of *Reconstruction* encompasses a conceptual movement where distinct philosophies are placed together often in surprising ways to provide a purview of world history. The complexity of the project accounts for the form and style of Iqbal's writing that some have suggested lacks conclusiveness.[7] Although the form resists the conclusiveness of empirical thought, it does not negate speculative truths.[8] In this form, seemingly incommensurable philosophical traditions are also placed adjacently to emphasize the coevality of Muslim and Western philosophical sources. Iqbal places Western sources beside oral and religious sources—including his own poetry. The modern form of *Reconstruction* arguably developed out of thirteenth-century Muslim thinkers and other poet-philosophers—notably Ibn 'Arabi (1165–1240)—who bridged disciplines and often incorporated their own poetic insights into their work.

Rather than documenting private experiences, Iqbal's popular and enduring poetry serves to confirm knowledge that is corporeal and historical and provides validation for the speculative truth-claims of his philosophy. The poet-philosopher did not see Muslim thought as totalizing knowledge, but as intrinsically dialectical and open-ended, requiring continuous historical interpretation—and much like Hegel—he foregrounded the reader's *experience*: "The Ultimate Reality, according to the Quran, is spiritual, and its life consists in its temporal activity" (155).

British idealism deeply shaped the form of Iqbal's thought. While he critiqued these thinkers from the turn of the twentieth century to the interwar period, he also contributed to their major insights. In Iqbal scholarship, Muslim modernist, socialist, and nationalist (Indian and Pakistani) views are often at odds; yet Muslim democratic humanism arguably *connects* these disparate positions without necessarily reconciling them. This is precisely why we use Adorno's idea of constellation to describe both the speculative dimension and form of *Reconstruction*. The limited Iqbal scholarship on the Western influences on his thought tend to focus on a single thinker, often overemphasizing them to the detriment of others. Thinking about Iqbal in constellation allows for a nuanced understanding of interrelated interpretive traditions and the unpacking of both 'mutual harmonies' and antimonies between them.[9] This means reading Iqbal's influences in relation to the 'whole'—in other words, as he would have read them.

Iqbal's Hegelian view of the whole concurs with the British idealist J.A. Smith who maintains "that the whole of what is *Real* is constantly in movement. It is a process of change. There is no static or inert background against which this process takes place" (qtd. in Boucher and Vincent 18). Echoing Smith, Iqbal writes:

> Thought is therefore the whole in its dynamic self-expression, appearing to the temporal vision as a series of definite specifications which cannot be understood expect by reciprocal reference. Their meaning lies not in their self-identity, but in the larger whole of which they are specific aspects.
>
> (*Reconstruction* 6)

Although Iqbal gleaned the racial basis of nationalism in Ernest Renan's (1823–1892) writing—and eventually witnessed them manifested in National Socialism—he also sought a Muslim state in a federated India, and even suggested a common army to discourage Indian Muslim nationalism. According to Renan, Islam was Europe's other due to its apparent epistemological inability to conform to the concept of the nation, which he defined by a common history and the collective desire to create a future from a shared past. Islam thus loses all of its particularity; in response, Iqbal's view of Islam is similar to the Hegelian *Geist*: an active and purposive unfolding of spirit in history—a temporal activity—as he states in his famous 1930 Presidential Address: "To Islam, matter is spirit realising itself in Space and Time" (*Thoughts and Reflections* 163).

Hegel and Nietzsche were important to Iqbal for different reasons. As early as *Development* (1908), Iqbal draws on Hegel to understand shifts in consciousness as mediated by social and historical conditions. Like Hegel, Iqbal sees the development of the self in relation to the unfolding spirit, with social conditions precluding certain varieties of human experience. Iqbal returns to early Muslim thought to examine how its decadence manifested conceptually in what he perceived to be the "bare universality" (xi) of Persian thought, along with its lack of a system. This concern with system was a constant theme from *Development* (1908) to *Reconstruction* (1930). The Hegelian 'system' attracted him in this respect:

> There is a system of continuity of this passage. Its various stages, in spite of the apparently abrupt changes in our evaluation of things, are organically related to one another . . . The world process, or the movement of the universe in time, is certainly devoid of purpose, if by purpose we mean a foreseen end—a far-off fixed destination to which the whole creation moves.
>
> (*Reconstruction* 54–55)

Yet Iqbal scholars emphasize or allude to the Nietzschean influence, while often overlooking—or deemphasizing—the Hegelian dimension. For example, Sevea writes: "Crucially, Iqbal's engagement with the works of western figures who examined and critiqued the modern condition remains unexplored. Nietzsche, of whose work Iqbal claimed to be unaware before his departure to Europe, is particularly important in this context" (20). Nietzsche appears in Iqbal's major Persian book of poetry, *Javid-Nama* (Book of Eternity), as do other important thinkers such as Tolstoy. However, Sevea and others downplay the poet-philosopher's own critique of Nietzsche, as well as how the latter's ideas played out in prominent Nietzschean philosophers, such as Spengler—frequently cited in Iqbal's work.

Although postcolonial and decolonial scholars consider Nietzsche a major philosopher who anticipated psychoanalysis and existentialism, his unappealing politics are glossed over, as well as his global reach.[10] While it is true that Nietzsche developed a distinctly literary register for philosophizing, the contradictory shifts in tone and position in his writing anticipate the postwar emphasis on ambiguity *as* politics (Brennan *Marxism*). In his own words, Nietzsche was the last apolitical German philosopher. However, few of these scholars speak about his undermining political philosophies such as socialism, social democracy, communism, and other 'herd' values that he was determined to overcome with new values.

Nietzsche aligns with cosmopolitan rootlessness and the dissolution of the nation-state, which are key themes in postcolonial and world literature. Since Iqbal does not clearly fit into socialism, communism, and nationalism, Nietzsche seems to provide an obvious interpretive fit to conceptualize Iqbal's work.[11] However, the opposite is true. Iqbal lamented the British presses' association of his ideas with those of Nietzsche. He often referred to Nietzsche's *politics*, and understood how the latter's ideas were playing out in interwar Europe—perhaps even in British eugenics writers such as Oscar Levy—whose Nietzsche translations he cited. On the other hand, numerous scholars document Iqbal's relationship with socialist ideas.[12]

The poet-philosopher's scathing critiques of capitalism and imperialism are evident in his poetry where he speaks of the unequal relationship between the worker and the capitalist; he considers Islam to be a socialistic religion as late as 1930. His family did not own land, and he criticized the Punjab Unionist party for its ideological support of communal landlords, which was a major influence from 1923 to the 1940s (Sevea 18). Although political economy is not a major strand in *Reconstruction*, Iqbal developed the first Urdu work on the subject. While commentators note his affinity and lack thereof to socialism, they overlook his anticolonial humanism, where Hegel is key in a way that reveals an important linkage with Vico. Notably, in addition to resituating Hegel in Edwardian Britain, British idealists were reinterpreting Vico—particularly the Neapolitan's idea that the natural world is unintelligible to us since God is its author; however, we can know the social world since we have authored it.[13]

Vico understood oppression to be national and international and anticipated the connection between class conflict and foreign domination so vital to anti-imperialism; and, as such, the philologist provides a powerful nexus to twentieth-century anticolonial and socialist thought (Brennan *Borrowed Light*). Vico's *New Science* also anticipates Hegel's Master-Slave dialectic, but the philologist does not hierarchize religion in the way that Hegel does in his *Lectures on the Philosophy of Religion*. Instead he understands the historical value of popular traditions as having "a public basis in truth" (Vico 81) and develops a hermeneutics that recuperates the vernacular and orature as rational social forms, which are of significant importance where such quotidian forms persist in the public sphere and serve as possible sites of anticolonial resistance. This is particularly the case in Iqbal's anticolonial humanism where religion and oral culture form a dialectic that puts into question contemporary neo-religious criticism with its emphatic critique of secularism.[14]

Vico railed against national isolationism: "Nations that develop in isolation allow them to overestimate their cultural productions" (xvii). Iqbal similarly seeks to develop 'mutual harmonies' between Muslim and Western thought and saw these harmonies as particularly evident in Western humanism that was borrowed from the Muslim world. In *Reconstruction*, he returns Muslim thought to itself, as it were, by placing it in relationship with Western thinkers such as Hegel, Bergson, and Nietzsche who—as he saw it—had advanced more than many contemporaneous Muslims who had lost touch with the *Geist*. Thus, he sees the spirit as worldly—and in much the same way that Goethe drew on Hafiz to rethink German literature, Iqbal pools together the best conceptual resources of German (and other) philosophy to rethink Islam during a critical moment of decline. Among the factors that influenced this were the decline of the Ottoman Empire, the British occupation of Egypt (1882), the Young Turk Revolution (1908), and the dearth of Indian Muslims hired into the civil service (Ahmad; Bayly; Hobsbawm).

Similar to Vico, Iqbal was a polymath and bridged fields including natural science, which also distinguishes his thought from the Neapolitan, whose *New Science* was the philosophical knowledge of the civic world and did not include natural sciences. Yet both thinkers provide polemical critiques of Descartes. Oral culture and religion are central to both thinkers and evident in every aspect of civic life. Iqbal, like Vico, does not separate rationality and sensation. Both philosophers critique scientism on the one hand, and Platonic thought on the other—as Iqbal writes: "Life is, then, a unique phenomena and the concept of mechanism is inadequate for its analysis" (44). Broadly speaking this reverberates with the British idealist critique of empiricism and utilitarianism.

Vico provides a seminal critique of the classical notion that the Greeks are divine in origin, in contrast to the plebeians who were bereft of Gods. Iqbal similarly criticizes Plato who "despised sense-perception" (*Reconstruction* 4). Vico writes: "countless abstract expressions, which permeate our languages today have divorced our civilized thought from our senses" (378). In the same way, Iqbal suggests purely rational readings of the Qur'an obscure its temporal poetics, which he sees as having rational content. Knowledge for Iqbal is "sense-perception elaborated by understanding" (*Reconstruction* 12). Such an anticlassical reading contests Muslim followers of Plato who created an opposition between poetry and thought, which developed into epistemological systems that formed the basis of juridical, theological, and grammatical systems (Adonis 57). Vico suggests that knowledge derives from sensation, which the intellect draws on in "an act of gathering" (136). Iqbal similarly sees sense-perception as an organ for historical knowledge that resonates with Vico's idea "that the poets were the sense of mankind, and the philosophers its intellect" (136). Iqbal's well-known aphorism, "Nations are born in the hearts of Poets; they prosper and die in the hands of politicians," surely would have resonated with Vico (*Thoughts and Reflections* 77). Less known is the aphorism written around the time of the Russian Revolution: "Philosophy ages; Poetry rejuvenates" (*Thoughts and Reflections* 92). Iqbal's placement of poetry and philosophy in critical relationship in the form of his political philosophy, and also his essays, have a broader purpose in Muslim thought. He often aligns reason with philosophy while sensation and intuition are correlated with poetics. In *Reconstruction*, Iqbal bridges reason and intuition and mediates two opposing ideas in Al-Ghazali's (1058–1111) thought. As Mir explains, one of Iqbal's major philosophical achievements for Muslim thought is in positing that intuition is not static:

> To this day, many Muslim intellectuals think that Al-Ghazali won the day for Islam by positing a special—i.e. mystic mode of consciousness that is distinct from the rational mode and is proof against rationalist doubt or attack. Iqbal rejects this idea by arguing, first, that religion itself is not without a strong rational element, and second, that no watertight division between thought and intuition can be made.
>
> (99)

Similar to British idealists, Iqbal rereads Kant's philosophy in light of Hegel, but he views intuition as historically mediated sensuous experience. This registers in the form of Iqbal's prose where poetry is often the language for non-perceptual modes of experience; and philosophy conceptualizes the latter. Both philosophy and poetry are integral to the *form* of Iqbal's prose—as they were to such British idealists as Bosanquet who understood poetry as the abstract idea clothed in sensual form (Hegel *Aesthetics*), as Mander writes:

> Few philosophers today would define their work in relation to poetry, but to the Idealists this was a natural and important relation for, as they saw it, there exists a deep coincidence of *aim* between the ends pursued by these two endeavours.
>
> (340)

As we will see, both are also important in Iqbal's conception of time as consisting of non-serial (experiential) time and serial (historical) temporality, which are key terms in his philosophy of history and opposed to Nietzschean eternal recurrence. Such a non-presentist conception of time resonates with Vico's cyclical conception of time that spirals through distinct ages. How did this tactile conception of thinking manifest in Iqbal's understanding of humanism, as compared to key British idealists?

In contrast to the British idealist, Bosanquet (1848–1923), Iqbal acknowledges that humanistic philosophy should start from 'plebeian material' rather than abstract ideals. Bosanquet proposed an ideological conception of the human, one that connected to the emerging League of Nations. In his hope to end all wars, Hobhouse (1918) famously challenged Bosanquet for not including other 'civilized' nations (106). Colonized states were not included in Hobhouse's critique because many British idealists adhered to social evolutionary ideas in regard to the colonies believing that 'inferior races' needed to develop to become capable of self-determination, although the Boer War (1899–1902) would prove to be the 'test' of their generation (Hobhouse; Mander; Boucher and Vincent). Despite these contradictions, Iqbal locates the actuality of a renewed humanist philosophy in the dialectical fabric of religion and oral culture in India.

His critique of the universal pretentions of British idealist humanitarianism faintly echoes (and obviously reinterprets) Marx and Engel's *The German Ideology*:

The humanitarian ideal is always universal in poetry and philosophy, but if you make it an effective ideal and work it out in actual life, you must start, not with poetry and philosophy, but with society exclusive in the sense of having a creed and well-defined.

(Iqbal qtd. in Hasan)

Far from being an advocate for nationalism in later life, Iqbal associates the unfoldment of Spirit with Islamic humanism: "At the present moment the national idea is racializing the outlook of Muslims, and thus materially counteracting the humanizing work of Islam" (2). He envisioned a future Islamic Studies as inaugurating a new humanism in the world (*Thoughts and Reflections* 103).

Iqbal's humanism begins from plebeian forms that he draws from popular traditions; thus, religion is not ideology, but part of the actual material from which social change develops and is restricted. In this he follows Bradley who in *Ethical Studies* (1876) attempts to rescue religion's separation from philosophy:

In order to be, religion must do. Its practice is the realization of the ideal in me and in the world. Separate religion from the real world, and you find it has nothing left it to do; it becomes a form, and so ceases. The practical content which religion carries out comes from the state, society, art, and science.

(166)

At a time of unprecedented imperial dominance, Bradley draws on Hegelian thinking that also becomes "eternally resolved" (174) so that philosophy becomes religious again.[15] In opposition to Bradley, Iqbal suggests that there is no finality to philosophical thinking, which echoes the social philosopher, J.S. Mackenzie's (1860–1935), language:

> Philosophy, indeed, may be compared to a serpent of eternity. It has neither beginning, middle, nor end. Every road in it leads us to the end of the world; and there is often as much hope of penetrating into the heart of the subject from the point that seems last as from the point that seems first.
>
> (3)

Similarly, in the Preface of *Reconstruction*, Iqbal writes: "there is no such thing as finality in philosophical thinkings [*sic*]. As knowledge advances and fresh avenues of thought are opened, other views, and probably sounder views than those set forth in these lectures, are possible" (vi).

Iqbal's philosophical view of self is in accordance with Hegel's self: "*to bring itself and indeed only its concept to appearance, to manifest its concept*" (Hegel qtd. in Pippin 41). The idea of the self as dynamic unfoldment is also emphasized by Oakeshott who states that "[a] thing is what it does . . . A thing that does nothing *is* nothing. Strip an object of its activities, and see what remains: You will find nothing" (qtd. in Boucher and Vincent 20). Such views were instrumental to Iqbal's Hegelian concept of self as a temporal activity that simultaneously manifests its concept. He sees the self as a dynamic and temporal activity and not the mere effect of the absolute. Thus, it would hardly be possible to imagine Iqbal without Hegel's looming influence, while also underscoring that Iqbal *reinterprets* the German philosopher through British Neo-Hegelians such as McTaggart, Bradley, J.S. Mackenzie, and other idealist philosophers. Raschid recognizes the centrality of Hegel in Iqbal's work: "The Hegelian influence is unmistakable and it is remarkable that Hegel is not explicitly mentioned" (8).[16] Iqbal likely drew much of his reading of Hegel from British idealists such as McTaggart, as he still refers to the triadic dialectic: thesis, antithesis, and synthesis—Fiche's appropriation of Hegel and not the philosopher's own rendering.[17]

Nonetheless, if the British idealists interpreted Hegel to challenge the utilitarianism of previous generations, Iqbal did the same in the Indian context. Bradley writes "[t]he mere individual is a delusion of theory" (173). The British idealists critiqued utilitarian philosophers such as Mill who understood Manichaeism as a "form of belief in the supernatural which stands wholly clear of both intellectual contradiction and moral obliquity" (116). Iqbal resituates this critique in the Indian context; his political philosophy similarly emerges out of a Hegelian critique of Khan who argues that a Western classical education was more important than Islamic Studies. Although British idealists were concerned with metaphysics, they also advanced educational reform—another driving theme of Khan and Iqbal's work. They sought to uplift the impoverished British working class mainly through education: "poverty for some [British idealists] represented moral failure of the individual" (Boucher and Vincent 115). Their interest in education—even if it was with the imperialistic intention to bring colonized subjects to the point where they were capable of self-determination—must have also nonetheless resonated with Iqbal who was concerned about the same issues in the Indian context.

While historians have commented on the impact of the Boer War on British anticolonialism and British idealism, few have speculated on its importance on Iqbal's intellectual formation and his critique of nationalism.

Iqbal in Britain after the Second Boer War

After the Second Boer War, a rapacious economic self-interest undermined Britain's image in the world and revealed crevices in British idealism. Hobsbawm states that although the Communist International signalled the beginning of secular left radicals' organizing around the banner of anti-imperialism, the period from the 1880s to 1914—with the rare exception—did not result in anti-imperialist political mobilization. However, most British labour leaders opposed the Boer War and advanced critical positions in regard to Britain's colonial holdings in Egypt, India, and Ireland. For instance, political activists exposed the harrowing conditions in the Congo and in urban plantations on African islands (Schneer; Hobsbawm). The 1906 electoral victory depended on the liberal party's condemnations of 'Chinese slavery' in South African mines (Hobsbawm 72).[18]

After 1894 the number of Muslim travellers arriving in imperial London rapidly increased as result of the Muslim intelligentsia developing a more receptive outlook towards British education (Lahiri 6). Iqbal arrived at Cambridge after the Second Boer War to study with the British idealist, J.M.E. McTaggart, at a time when Indians were viewed as a threat to the national security of the imperial state amid interracial fears, rather than an exotic curiosity—as they were perceived a generation earlier (Lahiri). During this period "the most studious [were] felt to be the most disloyal" (Lahiri 125). The Second Boer War was important in this policy shift and had far-reaching effects on Britain's image in the world and also on Iqbal's experience of racism.

Prior to Iqbal's arrival in London, the city hosted the first Pan-African conference in London in 1900 where W.E.B. Du Bois prophetically stated that "the problem of the twentieth century is the problem of the color-line" and spoke of the need for unity among colonized nations (19). British idealist philosophers were attempting to understand the metaphysical relationship between the individual and the state; yet the Second Boer War undermined their nationalism. The publication of J.A. Hobson's *Imperialism: A Study* (1902) was a culminating moment of a burgeoning European anticolonial internationalism.[19] Hobson's work—although deeply imbricated by racial assumptions—had a far-reaching world-historical impact on Lenin who was exiled in London around the same time as Iqbal was visiting.[20]

Iqbal's experience of racism in Britain was profoundly important for his intellectual formation, as he wrote in a letter: "I reacted strongly to this development; it brought about a revolutionary change in my whole outlook. In fact it was this atmosphere in Europe which turned me into a Mussalman" (Iqbal qtd. in Zakaria 31). Despite Iqbal's experience of racism and increasing religiosity, during this time he was sympathetic to the Indian Congress. While at Cambridge, he wrote about the 1905 Partition of Bengal and ensuing religious riots that marked a clear juncture in the divide and rule policy—further consolidating a separate Muslim identity. In order to represent these separate interests, the All-India Muslim League formed in 1906. Practically speaking during this period, the political intentions of the Muslim League and the Indian Congress more than often converged.

Similar to Du Bois's call for religious unity, Iqbal supported the Swadeshi movement, while also lamenting the lack of economic thought among Congress leaders, which perhaps suggests a familiarity with Hobson's work, published a few years prior to his arrival in Britain and discussed in British idealist circles. He writes:

> it is a pity that while our leaders agitate for political independence, they pay little heed to the basic need of our people, which is economic; it is prerequisite to the achievement of our independence. Fortunately, this aspect has begun to receive interest now. Likewise, there must be unity of objectives among different groups and individuals, constituting the nation; otherwise the inexorable law of nature will prevail and frustrate all their efforts to be united. In the result they will be wiped out. Our tragedy is that while we talk of unity, we do nothing to unite our people; it is the need of the hour.
>
> (qtd. in Zakaria 24)

Iqbal forged friendships with British idealists and immersed himself in the study of German philosophy, particularly Nietzsche—abandoning poetry. In addition, he also developed an interest in the Nietzschean *Übermensch*, which was in vogue during the Boer War. The Boer War impacted British idealists, some of whom echoed Hobson's critique of rampant capitalism. Edward Caird, for example, resisted the Boer War and protested against those who wanted to award Cecil Rhodes an Oxford degree (Vincent and Plant 89). Perhaps such moments allowed Iqbal to forge solidarities with the critical liberal thinkers of the time. As compared to the British Nietzscheans of the time who endorsed eugenics, the British idealists were liberal imperialists, or simply silent on the colonial issue.[21] Putting forward a subtle and penetrating comparison between Nietzschean politics and religious pity, Horkheimer and Adorno argue that the latter sharply contrasts with the fascist undertones of the Nietzschean "vilification of pity into an assent towards martial law" (*Dialectic* 80). That is, in comparison to Nietzsche's "vilification of pity," pity has a moment of truth—even as it attempts to negate and appease the conscience through guilt.

For some British idealists, philosophy was the pinnacle of experience and the manifestation of the absolute that resolved all contradictions. Although Iqbal upheld McTaggart as a "philosopher-saint," he also challenged his mystical view of the immortality of the self and contended that "the absolute of the neo-Hegelian lacks life and movement" (*Thoughts and Reflections* 124). However, his criticism is Hegelian in his affirmation of a relational understanding of the self: "The real test of a Self is whether it responds to the call of another Self" (126). Iqbal's concept of self is dynamic and not merely an effect of the absolute, or its substance.

In spite of his admiration for McTaggart, Iqbal revolted against the way he equivocated the Hegelian absolute with the immortality of the self. Instead, he viewed this as evidence of the static nature of McTaggart's thought, which he attributed to a parochial view of the world, and a prelude to the colossal loss of human life during the First World War: "McTaggart's view must be counteracted . . . by taking immortality as a hope, an inspiration, a duty and not as an eternal fact" (151). McTaggart's self should be "counteracted" and returned to the world and history, as Iqbal writes in a poem that was included in his essay: "I am very much worried about the loneliness of my 'Self'. Therefore, I am making arrangements of its *Society*!" (*Thoughts and Reflections* 151). This is reminiscent of Adorno's view that Hegelian being "negat[es] its absoluteness" (*Hegel* 32). In sum, Iqbal puts McTaggart's mystical categories into motion. The poet-philosopher also translates his ideas to a British audience and sharply contrasts them with Spengler's *The Decline of the West* (1918–1923), which was popular in interwar Europe. According to Iqbal, Spengler "seems to think that Islam amounts to a complete negation of the ego" (*Reconstruction* 109). For Iqbal, Spengler renders Islam into a passive object of control; and his critique of this view is in accordance with Adorno's statement that: "[Spengler's] entire image of history is measured by the ideal of domination" (*Prisms* 60)—and thus Nietzschean.

If Iqbal became a Muslim as a result of his experience of racism in Britain, this new outlook emerged because he saw Islam and humanism as intertwined. As he saw it, Islam's world-historical achievement was its non-racial humanist view. *Contra* Hegel, who did not view Islam as a determinate religion, Iqbal's *Reconstruction* situates Islam in human history and philosophy. The poet-philosopher sought a non-racial basis for humanism in the thought of the great Persian poet, Jalāl ad-Dīn Rumi (1207–1273), who he saw as on "opposite poles of thought" to Nietzsche (*Reconstruction* 89). He even approvingly cites Hegel's nod to the "great Rumi." Accordingly, his reading of the Qur'an emphasized the absolute's life as a temporal activity in the self—particularly manifested in the concept of love. As Iqbal sees it, the problem with Sufism is that it made the category of love into a Platonic ideal: "God's life is self-revelation, not the pursuit of an ideal to be reached" (*Reconstruction* 60). Sufism is in fact "a form of free thought and in alliance with Rationalism" (*Reconstruction* 150). The poet-philosopher gives loves a dialectical character by bridging the self and political community through this concept and also the appreciative and efficient dimensions of self. Iqbal develops his idea of recognition of the other by drawing on Rumi, but with a penetrating glance to Hegel. The affiliation between Hegel and Rumi makes sense if we recall Iqbal's idea that what is foreign resonates only if it is already pre-existing within the cultural fabric where it is grafted. Iqbal cites Rumi's poetry in different instances in *Reconstruction* to demonstrate his continued resonance for the Muslim world as a historical correlate to Hegel for Muslims. In short, the poet-philosopher searched for a nexus point between Muslim thought and Hegel's philosophy of history.

Rumi's poetics also served as a corporeal basis for politics in his connection of intuition with sensation and gives a conceptual form to the nonconceptual.

In contrast to the British Neo-Hegelians, Iqbal's philosophy of time seeks to historicize both conceptual and non-perceptual historical material. Similar to a materialist conception, Iqbal bridges time experienced with historical time (or serial time) and sees the past as carried within the present. This tension between serial and pure time generates temporality as becoming rather than Nietzschean eternal recurrence. Rumi's evolutionary concept of love is a response, or radical solution to Iqbal's reading of Nietzsche—since corporeality and politics converged in the Holocaust camp. Rumi's concept of love is both biological and social and would likely be the only antidote to counter the hardness of Nietzschean will-to-power, as Zarathustra asserts: "All creators, however, are hard" (113).

As compared to British idealists, Iqbal was far ahead of his time, particularly because of his international sensibilities and his dialectical humanism. Hobsbawm states that "internationally socialism before 1914 remained overwhelmingly a movement of Europeans and white emigrants or their descendants" (72). A global political awareness began to pervade the era, which became a consistent theme in Iqbal's poetry. In this respect, the Russian Revolution was the second juncture in Iqbal's understanding of Muslims as anticolonial subjects in his philosophy of history. In fact, several months after the Russian Revolution, Iqbal was grappling with Hegel and Nietzsche and casting the latter in an increasingly hostile light.

Iqbal, Nietzsche, and the Russian Revolution

Iqbal's fellow student and confidant, the travel writer Atiya Begum, recounts how the poet-philosopher started reading and discussing Nietzsche in Britain in 1907 with a group of other students and professors. The same year that he began his study of Nietzsche, he "prophe[sized] that Europe was ready to commit suicide with its own weapons, because she had converted the abode of mankind into a shop" (Fyzee-Rahamin xxiii). Iqbal's reading of Nietzsche was arguably instrumental to his insight. In his 1924 poetical work, *Bāng-i-Darā* (The Sound of the Caravan Bell), he writes that Europe would succumb to its death by its own dagger. The dagger alludes to a Nietzschean theme and can perhaps be read as a metaphor for the philosopher.[22] In the Nazi camps of Europe and in the colonies, the political forces of death and human depravity eventually did appropriate the *Übermensch*: this is precisely why Iqbal sees in Nietzsche a prophet of decay.

The poet-philosopher took issue with Nietzsche's idea of will-to-power. However, he also saw Nietzsche's importance in foreseeing the world situation: if life is will-to-power, this view would result in the rise of Western fascism and the emergence of the modern death camps. Thus, Iqbal's reading of Nietzsche as prophet of decay echoes Adorno and Horkheimer's insight about the *Übermensch*: "By elevating the cult of strength to a world-historical doctrine, German fascism took it to its absurd conclusion" (*Dialectic* 79). Iqbal sees Nietzsche's will-to-power as on the side of the "superior man who has been robbed of power and . . . should assert himself and tell the inferior to remain where they should be" (242). This is reminiscent of Vico's understanding of the nobles' proud avarice and barbarous actions towards the plebeians.

On the one hand, Nietzsche's famous critique of the freedom of will challenges its underlying motivation as

> an affect, and specifically the affect of the command. That which is termed "freedom of will" is essentially the affect of superiority to him who must obey . . . A man who *wills* commands something within himself that renders obedience, or that he believes renders obedience.
>
> (*Beyond Good and Evil* 25)

The qualification "believes" is important; on the other hand, Nietzsche stresses the will as domination—not the will of the oppressed to resist domination and create their history. If humanism masks barbarism and the freedom of will conceals the impulse to dominate weaker wills, than philosophy and history are bereft of meaning. Hence Iqbal contends, "to [Nietzsche] there is no ethical principle resident in the forces of history" (*Thoughts and Reflections* 242).

While the critique of freedom of will was important for Nietzsche, it was not likely relevant in colonized states where free will was an imperial privilege. Hegel's view of freedom was the outcome of a collective struggle: "freedom itself, must be understood as a kind of social achievement in time, a result of opposition and contention over such a 'precognitive' status" (Pippin 35). This was certainly a concern for British idealist social philosophers such as J.S. Mackenzie, although these philosophers limited their discussions of class disparity to the national context. Thus while the *Übermensch* wields the will-to-power as a political weapon, Iqbal attributes the concept of nobility to the plebeian rather than the *Übermensch*; in this, we can hear Vico—as Iqbal writes in April 1917:

> The fate of the world has been principally decided by minorities. The history of Europe bears ample testimony to the truth of this proposition. It seems to me that there are psychological reasons why minorities should have been a powerful factor in the history of mankind. Character is the invisible force which determines the destinies of nations, and an intense character is not possible in a majority.
>
> (*Thoughts and Reflections* 79)

In sharp contrast with Nietzsche's searing opposition to the democratic intermingling of class and races in Europe, Iqbal instead argues "the history of Europe bears witness to the fact that the less dominant minority determines the progress of history" (79). According to Nietzsche's view, character results in the recurrence of experience, and thus resembles a religious conception of time (Benjamin). Along the same lines, Iqbal suggests that Nietzsche did not address the problem of time "and accepted without criticism the old Hindu and Greek idea of time" (*Thoughts and Reflections* 241–242). For Iqbal, character develops in the oppressed minorities who, as we will see, reinterpret popular traditions. In his critique, Iqbal challenges Nietzsche's concept of character and gives it dialectical significance by exposing its contradictions and inverting it.

Notably, Iqbal's critique of Nietzsche became acerbic in a remarkable essay entitled 'Stray Thoughts' (1917) where he mounted a critique of Nietzsche—opposing his ideas with a Muslim democracy—which he aligned with a kind of socialism. The tensions between Nietzschean *Übermensch* and the plebeian become clear in this essay. It is the 'plebeian material' within a democratic humanist Islam that is important, rather than the *Übermensch*:

> Nietzsche . . . abhors this "rule of the herd" and, hopeless of the plebeian, he bases all higher culture on the cultivation and growth of an Aristocracy of Supermen. But is the plebeian so absolutely hopeless? . . . Out of the plebeian material Islam has formed men of the noblest type of life and Power. Is not, then, the Democracy of early Islam an experiential refutation of the ideas of Nietzsche?
>
> (*Thoughts and Reflections* 83–84)

It would of course be a mistake to view Iqbal's plebeian as solely Muslim: "all men and not Muslims alone are meant for the Kingdom of God on earth, provided they say good-bye to their idols of race and nationality, and treat one another as personalities" (*Thoughts and Reflections* 99). Rather, we suggest plebeian material as not identical solely with Islam, but rather those historical elements that are important for the development of democratic institutions. We also suggest that Iqbal's use of Nietzschean concepts such as 'power' should be understood in relation to his immanent critique of the philosopher. These categories are not exploded from within, but rather transfigured by intrinsically drawing on their own critical force to immanently develop them towards anticolonial ends.

If, by 1911, Iqbal had "taken refuge in Nietzsche" (Begum 35), by 1917 he sees the rejuvenating importance of poetry and, in aphoristic brevity, refers to "Hegel's system of Philosophy [as] an epic poem in prose" (*Thoughts and Reflections* 91). Throughout his life, Iqbal seems to have struggled between poetry and philosophy, which he saw conjoined in Hegel's speculative form. This helps him to develop his conception of the self-efficient and appreciative self, which he respectively aligns with two parts of the self: realism and concrete hope.[23] Poetry is the form of concrete hope—emerging from plebeian material—that philosophy interprets and gives directive force and power. The recurring Qur'anic quote in *Reconstruction* supports this reading: "To Him belongs creation and direction" (103). Iqbal compellingly includes his own poems in his own essays and in *Reconstruction*.[24] In fact, *Reconstruction* ends with a selection from *Javid-Nama*—emphasizing the importance of creative experience in the interpretative act, which lends renewed directive force to his philosophical prose.

Echoing Trotsky's critique of pure art, Iqbal provides a powerful critique of *l'art pour l'art*.[25] Writing in July 2017, he states "there should be no opium-eating in Art. The dogma of Art for the sake of Art is a clever invention of decadence to cheat us out of life and power" (*Thoughts and Reflections* 86). While Nietzsche's Zarathustra descends from a mountain to impart the *Übermensch*'s wisdom to those below, Hegel's consciousness begins from the earth and ascends, gathering and developing the elements towards human freedom—as Hegel writes: "The History of the world is none other than the progress of the consciousness of Freedom" (*Lectures* 33).

As we suggested, Iqbal provides an intrinsic critique of Nietzsche's idea of character. For instance, he associates nobility with 'plebeian material', which is not incidentally determined by class or other origins—as he states:

> The fact that the higher emerges from the lower does not rob the higher of its worth and dignity. It is not the origin of a thing that matters, it is the capacity, the significance, and the final reach of the emergent that matters.
>
> (*Reconstruction* 106)

In this, Iqbal has proved to be remarkably prescient, as he emphasizes the movement of elements in a configuration, and thus his thought is not reducible to identity.

Iqbal wrote to his translator, Nicholson, in 1921 in response to a review of his poetry where he states that if a British reader wants to grasp his poetry they should refer to J.S. Mackenzie's plebeian view of the ideal man as "in the streets of our crowded city" (*Thoughts and Reflections* 96). He goes on to suggest that to understand his view of the ideal man, a British reader should refer to Mackenzie's British Neo-Hegelian work, *An Introduction to Social Philosophy* (1890): "It is in light of the above thoughts that I want the British public to read my description of the ideal man" (96). He contrasts this with the reviewer's comparison of his translated poetry to Nietzsche and goes on to rebuke of the ascetic aspects of the philosopher's idealism. Iqbal remarks that "[s]ome of the English reviewers, however, have been misled by the superficial resemblance of *some of my* ideas to those of Nietzsche" (91). According to Iqbal, Nietzsche's eternal recurrence is repetition of the same: "not eternal creation; it is eternal repetition" (142).

> Nietzsche, although he thought that the idea of evolution did not justify the belief that man was unsurpassable, cannot be regarded as an exception in this respect. His enthusiasm for the future of man ended in the doctrine of eternal recurrence—perhaps the most hopeless idea of immortality ever formed by man. This eternal repetition is not eternal "becoming"; it is the same old idea of "being" masquerading as "becoming."
>
> (187)

Although there are clear allusions to the Master-Slave dialectic in The Reconstruction, Iqbal does not describe a life-and-death struggle for self-determination, but provides a more psychological or Kantian rendering of Hegel: "I am distinct from and yet intimately related to that on which I depend on for my life and sustenance" (*Reconstruction* 118).[26] His attempt to reconceptualize the Hegelian subject through Kant also draws on British idealism. Hegel conceptualizes servitude through fear, which he sees as part of a process towards self-determination. The poet-philosopher similarly sees fear as an obstacle to freedom—rather than stressing Nietzsche's idea of domination—which aligns with the ruler. In Iqbal's Kantian rendering of Hegel, love is an activity of self that undermines the fear that prevents freedom. In Hegel's *The Phenomenology of Spirit*, fear closes off subjective development and renders consciousness mute and internalized.

Clearly, it is Hegel's vocabulary that Iqbal is drawing on when he states: "[whatever may be our view of the self—feeling, self-identity, soul, will—it can only be examined by the canons of thought which in its nature is relational, and all 'relations' involve contradictions" (*Reconstruction* 98). Iqbal, by contrast, viewed Nietzsche's *Übermensch* as an unattainable ideal—more aligned with the Vedantic idea of *maya*. While Nietzsche appears in Europe as a prophet of a distinctly modern philosophy, in India he appears as ancient. Nietzsche did not address the problem of time "and accepted without criticism the old Hindu and Greek idea of time" (*Thoughts and Reflections* 241–242).

In *Reconstruction*, Iqbal locates a similar idea in the Muslim view of *kismet*. The word comes from the Ottoman Turkish word *qismet* and *kismet* later implied fate, religiously conceived. The poet-philosopher was likewise extremely critical of Spengler's view of Islam as entailing a negation of the ego. Iqbal instead echoes the British idealist, J.S. Mackenzie, who begins with the 'ideal man' to think about 'ideal society', which resonates with Muslim culture, where great thinkers were often profoundly ordinary. This view also likely struck a chord with Iqbal, due to his own Punjabi background that irritated Urdu intellectuals (Sevea 18).

In order to understand Iqbal's view of the self, we also need to consider his critical reading of the French philosoper Henri Bergson, whom he met. His sympathetic (and dialectical) reading of Bergson helps us to understand his philosophy of history and distinguish it from the teleological thinking of some British idealists who—for the most part—rejected Bergson's anti-intellectualism (Mander 476). Through his unique philosophy of time, Iqbal challenges the teleological assumptions of many British idealists as well as those of Nietzsche.

Muslim modernity, plebeian poetics, and orature

In Iqbal's philosophy of history, he is continually attempting to align sensation with thought. Sensation for Iqbal is both physical and nonconceptual. For Iqbal, serial time is not a line with a predetermined beginning and endpoint, but exists in relation to pure time and unfolds within it: "A time-process cannot be conceived as a line already drawn. It is a line in the drawing—an actualization of open possibilities" (*Reconstruction* 55). Thus, when Iqbal refers to teleology, he does not mean that serial time is divinely preconceived and preordained, but rather that it exists in relation to the creative possibilities of the appreciative self, which interacts with the nonconceptual, placing intuition in relationship to history and preserving the sensuous dimension of reason. The appreciative self does not develop in isolation, but in relation to serial time. Iqbal thus resists Bergson's overemphasis on subjectivity in his philosophy of time—but not entirely. In Iqbal's conception of temporality, the movement of serial and non-serial time does not result in being and essence, but becoming that is not helplessly opposed to rationality as it is in Bergson (Adorno *Prisms*).

Unlike Bergson, Iqbal does not see serial time as opposed to pure time, which would result in the separation of the appreciative self from the self-efficient self. Iqbal suggests that "serial time is pure duration pulverized by thought— a kind of device by which Reality exposes its ceaseless creative activity to qualitative measurement" (*Reconstruction* 58). The poet-philosopher defines pure time in a non-presentist manner: "Pure time . . . [is] an organic whole in which the past is not left behind, but is moving along with, and operating in, the present" (49). Pure time opens up creative points of possibility of interruption within serial time.

The emergence of the Qur'an signals Muslim modernity as an interruption of serial time through the creative event. Pure time is thus the historically enabled subjective possibility to collect past thought and challenge the reproduction of serial time in both the psyche and the human community. The tension between serial and pure time is becoming temporality—rather than eternal recurrence. Subjectivity is not erased by predetermined categories in an onward march towards its destination, nor is it developed in a linear manner; rather, time cycles and the past reappears in new guises. Iqbal frequently quotes the Qur'anic idea of the alternation of day and night in *Reconstruction*, suggesting a cyclical view of history similar to Vico's *corsi e ricorsi*. Individuals create the present even as historical forces shape their character. Furthermore, Iqbal asserts that we do not experience temporality passively: "and the future is given . . . only in the sense that it is present in its nature as an open possibility" (49). He does not overemphasize free will in the way that Bergson did in his philosophy of time. Character is generated through an accretive movement that develops out of the interweaving of time as event and time as experience.

For Iqbal the idea of natural history is a congealed poetics that is evident in oral culture. As we have shown, intuition is not a natural category opposed to history, but like Benjamin, Iqbal recuperates the apparently natural elements of history, particularly in his poetry. Iqbal's idea of natural history[27] is evident in Iqbal's revision of the Sufi parable of the moth and the flame. In this story, the moth represents the human subjectivity that becomes burned and annihilated by the flame of eternity—a popular Sufi metaphor—reinterpreted by the Persian poet, Farīd ud-Dīn 'Aṭṭār (1145–1221). Iqbal's persona instead is tired of the moth and flame and wants to see the moth swallow the flame. In Iqbal's reinterpretation, the decaying parable comes to life again. The flame does not consume the moth; but rather aspiring humans contain within themselves the flame, while retaining their sense of self. His novel rendering of this parable alludes to a separation between the subject and object, which does not allow for synthesis. For Iqbal, *fana* (sufi annihilation)—at least in the form in which it manifested during his time—was an effect of colonization, a symptom of decline where Muslim intellectual visionaries retreated from politics. The moth does not become the flame; rather the subject and object retain a necessary separation that enables movement and criticism. In Iqbal's rendering, the subject and the object retain a critical intimacy, but they are not identical; and thus they resemble Adorno's conception of the subject.

This short illustration helps us to grasp the relationship between poetry and Iqbal's philosophy of history, where the two interconnect and put the emphasis on orature and the vernacular as modern and rational forms that require philological interpretation. In short, Iqbal challenges the idea that the Qur'an is a singular event manifested in its concrete actuality of sacred words. The interpretation of words is read in light of the *hadiths*, oral traditions, modern thought, and his own poetry—or else the book becomes an archaic recurrence of the same and an instrument for domination and appropriation. Creative experience allows for reinterpretation within the book's own historical horizons, and philosophical interpretation gives it direction and material force. Again, Iqbal attempts to bridge the subject and object without sealing the intimate gap between them. He mines and creatively develops the oral traditions. Thus, creative destruction is not as important as imaginative development of the thought of the past.

Iqbal provides a humanist basis for Islam as a development of thought of the past. This view can be contrasted with Žižek's emphasis on the Real, which overlooks the power of the symbolic dimension in Islam and emphasizes historical rupture. As opposed to Christianity, Žižek suggests that in Islam there is no mediation of the Father (through the symbolic realm), who constitutes the community through guilt in Christianity. Thus, unlike Christianity, he suggests that Islam retains a *direct* and unmediated relationship with politics manifested in the Islamic state—and its possibility for co-option in the form of a singular interruption of history—perhaps in the form of anti-state collectives without any central organizing principle. The Prophet of Islam's human sexuality is not repressed or idealized. He is neither desexualized nor perverse, but profoundly human. His first mystical experience does not result in surrender, but trembling bewilderment.

If intuition for Iqbal possesses a relation to time, his rendering of it is paradoxical, as it is seemingly experienced a priori while also containing cognitive content. The philosopher-poet at times confuses metaphysical and historical categories; and yet metaphysical categories do also become historical.[28] In this reading, we have emphasized Iqbal as a speculative thinker within the Hegelian tradition. However, Iqbal's attempt to recuperate the historical dimension of the nonconceptual conflicts with his theory of the political power that stresses the domination of nature, which is a recurring theme in his work, for example: "Knowledge must begin with the concrete. It is the intellectual capture of and *power* over the concrete that makes it possible for the intellect of man to pass beyond the concrete" (*Reconstruction* 131). Thus, Iqbal's critique of instrumental reason conflicts with his stress on the domination of nature and the conquest of matter. This tension is not ultimately resolved in his work. When one takes recourse in Iqbal's concept of love, the latter stands helplessly before the domination of nature.

In Iqbal's 1938 New Year's speech on the All India Radio, he speaks in a remarkably prescient manner to our present moment:

> Today space and time are being annihilated and man is achieving amazing successes in unveiling the secrets of nature and harnessing its forces to his own service. But in spite of all these developments, tyranny of imperialism struts abroad, covering its face in the masks of Democracy, Nationalism, Communism, Fascism and heaven knows what else besides. Under these masks, in every corner of the earth, the spirit of freedom and the dignity of man are being trampled underfoot in a way of which not even the darkest period of human history provides a parallel.
>
> (*Speeches* 226)

It is tempting to read Iqbal's critique of democracy as Nietzschean and overlook the (Hegelian) spirit that often gets lost when situating him. In spite of Hegel's apparent ethnocentrism, the poet-philosopher sees himself as *reclaiming* Hegelian thought, particularly when he situates Islam as a determinate religion: "This constant contact with the spirit of Islam, as it unfolds itself in time, has, I think, given me a kind of insight into its significance as world fact" (*Thoughts and Reflections* 161).

When Hegal composed the Master-Slave dialectic, he was gesturing to the Haitian Revolution (Buck-Morss *Hegel and Haiti*). Anticolonial writers took Hegel's cue and stretched the dialectic to account for unique problems: to not recognize this amounts to the erasure of his influence in the postwar refashioning of intellectual history (Brennan *Marxism*).[29] In addition, Iqbal argues that communism's failure was a result of it "ris[ing] in revolt against the very [Hegelian] source which could have given it strength and purpose" (*Reconstruction* 188). This important statement echoes Adorno and Horkheimer's critique of Soviet communism. However, Iqbal also followed the developments that emerged out of the Communist International, which shaped his own positions in addition to his endorsement of the Marxian Progressive Writers' Movement. To overlook this connection, to read Iqbal in a single national context, or to focus on a single thinker belies the poet-philosopher's worldly anticolonial humanism.

Notes

1 British idealist scholars agree that Hegel was the defining figure—see Mander for a detailed summary of key influences. Karim Dharamsi also emphasized to me the importance of Kant and, to a lesser degree, Vico. For more on Kant's influence, see Mander (51–61).

2 Iqbal's democratic humanism provides a contrast to Nietzsche who is often emphasized in the limited postcolonial readings of Iqbal. For example, see Roy Jackson's *Nietzsche and Islam*, particularly pp. 57-62, Sevea p. 102, and *Muhammad Iqbal: Essays on the*

Reconstruction of Modern Muslim Thought. Remarkably, Iqbal is not even included in the index of Robert Young's influential *Postcolonialism: An Historical Introduction*. The poet-philosopher is entirely neglected in recent British idealist scholarship. There is a large body of Indian and Pakistani scholarship on Iqbal.

3 It is important to note that Iqbal rightly saw the humanist movement as "set free by Muslim thought" (*Thoughts and Reflections* 104). Thus, when we speak of borrowing, we emphasize that it occurred in *both* directions.

4 In this chapter, Edward Said's idea of beginnings is drawn on to suggest that Iqbal's use of Hegelian concepts reveals a common vocabulary and even a similar working through of ideas around the subject to other major anticolonial figures—for instance, one can detect a resonance with Frantz Fanon's "Algeria Unveiled" in *A Dying Colonialism*.

5 See Sevea (2012) and Majeed (2007).

6 See Ali Mirsepassi's *Political Islam, Iran, and the Enlightenment: Philosophies of Hope and Despair*, particularly "Heidegger and Iran: The Dark Side of Being and Belonging." Notably, Mirsepassi critiques the Hegelianism within Muslim modernists. However, I suggest that Iqbal was a Muslim Hegelian on the left who challenged imperialistic Hegelianism from within.

7 See Majeed who suggests that the form of *Reconstruction* lacks conclusiveness. However, in *Reconstruction* when Iqbal states that philosophical thought lacks finality, he is echoing the British idealist, J.S. Mackenzie's, Hegelian sentiment about philosophical thought. As such, I would argue that the form of his work is best understood in relation to such thought that emphasizes speculative and metaphysical truths rather than inconclusiveness. For another view, see Mir, who provides a closer reading of *Reconstruction*. What is inconclusive in Iqbal's thought is the blurring of speculative and metaphysical truths.

8 Brennan (*Borrowed Light*) aligns the montage form with peripheral writers who emphasized the vernacular and popular forms rather than with literary modernism. Relying on 'difference', Majeed reads the distinct philosophies in *Reconstruction* as incommensurable. However, particularly since Iqbal is drawing on Hegel, we see these traditions mediated, or in constellation.

9 Adorno's philosophy of history is not a predetermined category ontologically grounded in the concept of 'progress', 'spirit', or 'nature' (Buck-Morss *The Origin of Negative Dialectics*). It is not derived from a mover of history—whether that is the proletariat or class-consciousness. Adorno positions his philosophy of history in opposition to Lukács' understanding of the proletariat as the subject and object of history to create a space for thinking. For Adorno, there is an unbridgeable gap between the subject and object that problematizes philosophical knowledge of its object, but that also makes criticism possible. Adorno was attempting to construct a dialectics without a teleological idea of history (Buck-Morss *The Origin of Negative Dialectics*). To think in constellations means that there is not one core concept to which cultural phenomena refer, nor can theory be applied to cultural objects extrinsically. Cultural objects are themselves theoretical artefacts, repositories of social antagonisms that exist in their form beyond authorial intentionality. Concepts are not unmediated in the critic's consciousness a priori in a Kantian fashion, nor are they superstructural effects of the economic base. Adorno's idea of critique does not begin from generalizing principles that privilege a theoretical standpoint, or a transcendental 'ought' external to the cultural object examined, nor is the meaning of a work derived from first principles.

10 See Detwiler, Stone, and Brennan (*Borrowed Light*). Stone is particularly important as his work addresses British Nietzschean philosophers in Edwardian Britain. Notable in this context is Oscar Levy, as Iqbal read Levy's Nietzsche translations. Briefly, Levy was a German-Jew who believed in the overthrow of Judeo-Christian values, replacing them with aristocratic ideal values. Iqbal seems to have been familiar with these

writers, as his critiques emphasize "Nietzsche's politics." Nietzsche was of marginal importance to British idealists such as J.S. Mackenzie who was more concerned with class disparity in Britain. For more on the influence on Nietzsche on British idealism, see Mander (476-478). Their aversion to Nietzsche echoes Iqbal's.

11 For a reading of Iqbal that emphasizes cosmopolitanism, see, for example, Jalal in Bose, Sugata, and Manjapra. *Cosmopolitan Thought Zones: South Asia and the Global Circulation of Ideas*. Palgrave Macmillan, 2010.

12 For example, see Jan Marek's "Socialist Ideas in the Poetry of Mohammad Iqbal" in *Mohammad Iqbal*. Edited by Verinder Grover, Deep and Deep Publications, 1995.

13 See Boucher and Vincent (14). In addition, Mander suggests that Bosanquet and Caird understood poetry and philosophy to be manifestations of the same truth in different forms.

14 For the defining example, see Talal Asad's *Formations of the Secular: Christianity, Islam, Modernity*. Stanford UP, 2003. Mirsepassi rightly suggests that Asad's concept of Islam is "anchored in a notion of romantic authenticity" (8).

15 For more on the influence of Bradley and McTaggart, see Javed Majeed's "Putting God in His Place: Bradley, McTaggart, and Muhammad Iqbal." *Journal of Islamic Studies*, 4.2, 1993, pp. 208-236.

16 In fact, Hegel is mentioned in a number of places in *Reconstruction*. Raschid is concerned with Iqbal's lack of consistent application of Hegel. However, this chapter examines Iqbal's affiliation with Hegel as a 'beginning' for his anticolonial philosophy of history rather than his correct or incorrect application of Hegel.

17 McTaggart adheres to the thesis-antithesis-synthesis rendering of Hegel in *Studies in Hegelian Cosmology*—a work that Iqbal frequently cited.

18 See also Stephen Howe's *Anticolonialism in British Politics*, Clarendon Press, 1993, particularly the chapter "Socialism and Empire before 1939"; also see Gregory Claeys, *Imperial Skeptics: British Critics of Empire, 1850–1920*. Cambridge UP, 2010.

19 Cain's (2002) *Hobson and Imperialism*, Oxford UP, examines the evolution of Hobson's theory of imperialism.

20 As is well documented, Iqbal wrote a monumental poem, *Khuda Ke Hanzoor Mein* (Lenin Before God), within a longer poetic sequence of odes, ghazals etc. (*Bal-i-Jibril*). In the poem, Lenin polemically protests to God about the dire conditions of the poor. Iqbal allusively mobilizes Sufi tropes and gives them direction in relation to the "jolt" of the Russian Revolution.

21 See Stone (2002).

22 The dagger appears in numerous instances in Nietzsche's corpus. For one example, see *Human, All Too Human*: "Pity.—In the gilded sheath of pity there is sometimes stuck the dagger of envy."

23 The term "concrete hope" is from Ernst Bloch who opposes it to abstract, or utopian hope. Similar to Bloch, Iqbal also uses the concept "not-yet" to distinguish concrete hope from utopianism.

24 See Iqbal's 1932 essay "McTaggart's Philosophy" that is interspersed with his own poems. *Reconstruction* ends with Iqbal's poetry.

25 See Trotsky's *Literature and Revolution*, published in 1924.

26 However, there are clear references to slavery in Iqbal's poetry.

27 For more on Adorno's idea of natural history, see "The Idea of Natural History," as well as Hullot-Kentor's introduction to this work.

28 For instance, immortality becomes hope in Iqbal's critique of McTaggart.

29 Bayly argues for a revised Indian "worldly" historiography that situates South Asia in Britain and stresses British critics of empire such as humanists, socialists, as well as liberal critics (*Origins of Nationality* 276). On the other hand, not situating Iqbal within Britain risks ceding the contributions of British idealism—such as a burgeoning theory of the welfare state (see Vincent and Plant)—to Western thought.

Works cited

Adonis. *An Introduction to Arab Poetics*. University of Texas Press, 1990.

Adorno, Theodor. *Hegel: Three Studies*. Translated by Shierry Weber Nicholsen. MIT Press, 1994.

—. *Prisms*. Translated by Shierry Weber Nicholsen and Samuel Weber. MIT Press, 1981.

—. *Negative Dialectics*. Translated by E.B. Ashton. Continuum, 1983.

—. "The Idea of Natural History." *Things Beyond Resemblance: Collected Essays on Theodor W. Adorno*. By Robert Hullot-Kentor, Columbia UP, 2008, pp. 252–269.

Ahmad, Eqbal. *The Selected Writings of Eqbal Ahmad*. Columbia UP, 2006.

Al-Ghazzali, Imam. *Incoherence of the Philosophers*. Adam Publishers, 2007.

Bayly, Christopher. *Origins of Nationality in South Asia: Patriotism and Ethical Government in the Making of Modern India*. Oxford UP, 2001.

—. *The Birth of the Modern World: 1780–1914*. Blackwell Publishing, 2004.

Benjamin, Walter. "Fate and Character." *Reflections: Essays, Aphorisms, Autobiographical Writing*. Edited by Peter Demetz. Schocken Books, 1986, pp. 304–311.

Bloch, Ernst. *The Utopian Function of Art and Literature: Selected Essays*. MIT Press, 1988.

Boucher, David and Andrew Vincent. *British Idealism: A Guide for the Perplexed*. Bloomsbury Publishing, 2011.

Bradley, Francis. *Ethical Studies (Selected Essays)*. The Liberal Arts Press, 1951.

Brennan, Timothy. *Borrowed Light: Vico, Hegel, and the Colonies*. Stanford UP, 2014.

—. "Postcolonial Studies between the European Wars: An Intellectual History." *Marxism, Modernity and Postcolonial Studies*. Edited by Crystal Bartolovich and Neil Lazarus, Cambridge UP, 2002, pp. 185–203.

Buck-Morss, Susan. *Hegel, Haiti, and Universal History*. University of Pittsburgh Press, 2009.

—. *The Origin of Negative Dialectics: Theodor W. Adorno, Walter Benjamin and the Frankfurt Institute*. Harvester Press, 1977.

—. *Thinking Past Terror: Islamism and Critical Theory on the Left*. Verso, 2003.

Detwiler, Bruce. *Nietzsche and the Politics of Aristocratic Radicalism*. The University of Chicago Press, 1990.

Du Bois, William. *The Souls of Black Folk*. New American Library, 1903.

Durrani, Saeed. *Iqbal: Europe Mein*. Iqbal Academy, 1985.

Fyzee-Rahamin, Atiya Begum. *Iqbal*. Oxford UP, 2011.

Green, Ewen. *The Crisis of Conservatism: The Politics, Economics and Ideology of the Conservative Party, 1880–1914*. Routledge, 2005.

Hasan, Qamar. *Muslims in India: Attitudes, Adjustments, and Reactions*. Northern Book Centre, 1987.

Hegel, Georg. *Introductory Lectures on Aesthetics*. Translated by Bernard Bosanquet, edited by Michael Inwood, Penguin Books, 2004.

—. *Lectures on the Philosophy of Religion: One-Volume Edition, The Lectures of 1827*. Oxford UP, 2006.

—. *Phenomenology of Spirit*. Translated by A. Miller and J. Findlay. Clarendon, 1977.

— and John Sibree. *The Philosophy of History*. Courier Corporation, 2004.

Hobhouse, Leonard. *A Metaphysical Theory of the State: A Criticism*. George Allen & Unwin Ltd, 1960.

Hobsbawm, Eric. *Age of Extremes: The Short Twentieth Century, 1914–1991*. Abacus, 1995.

—. *The Age of Empire*. Weidenfeld & Nicolson, 1987.

Hobson, John. *Imperialism: A Study*. J. Pott, 1902.

—. *The Crisis of Liberalism: New Issues of Democracy*. Barnes & Noble, 1974.

Horkheimer, Max, and Theodor W. Adorno. *Dialectic of Enlightenment: Philosophical Fragments*. Ed. Gunzelin Schmid Noerr. Translated by Edmund Jephcott. Stanford UP, 2002.

Iqbal, Muhammad. *The Development of Metaphysics in Persia: A Contribution to the History of Muslim Philosophy*. Luzac & Company, 1908.

—. *Javid-Nama (Rle Iran B)*. Vol. 14. Routledge, 2011.

—. *The Reconstruction of Religious Thought in Islam*. Kitab Bhavan, 1974.

—. *Speeches and Statements of Iqbal*. Edited by A.R. Tariq, Sh. Ghulam Ali & Sons, 1973.

—. *Thoughts and Reflections of Iqbal*. Edited by Syed Abdul Vahid, SH. Muhammad Ashraf, 1964.

Iqbal, Muhammed. *Tulip in the Desert: A Selection of the Poetry of Muhammad Iqbal*. Translated by Mustansir Mir. McGill-Queen's Press, 2000.

Jackson, Roy. *Nietzsche and Islam*. Routledge, 2007.

Khan, Syed. *Writing and Speeches of Sir Syed Ahmad Khan*. Edited by Shan Mohammad, Nachiketa Publications Limited, 1972.

Lahiri, Shompa. *Indians in Britain: Anglo-Indian Encounters, Race and Identity 1880–1930*. Frank Cass Publishers, 2000.

Mackenzie, John Stuart. *An Introduction to Social Philosophy*. J. Maclehose & Sons, 1895.

Majeed, Javed. *Autobiography, Travel and Postnational Identity: Gandhi, Nehru and Iqbal*. Palgrave Macmillan, 2007.

Mander, William J. *British Idealism: A History*. Oxford UP, 2011.

McTaggart, John. *Studies in Hegelian Cosmology*. Garland Publishing, Inc., 1984.

Mignolo, Walter. *The Darker Side of Western Modernity: Global Futures, Decolonial Options*. Duke UP, 2011.

Mill, John. *Three Essays on Religion*. H. Holt, 1878.

Mir, Mustansir. *Iqbal*. Oxford UP, 2006.

Mirsepassi, Ali. *Political Islam, Iran, and the Enlightenment: Philosophies of Hope and Despair*. Cambridge UP, 2010.

Nietzsche, Friedrich. *Beyond Good and Evil: Prelude to a Philosophy of the Future*. Translated and edited by Walter Kaufmann, Vintage Books, 1966.

—. *Human, All Too Human: A Book for Free Spirits*. Translated by R.J. Hollingdale, Cambridge UP, 1996.

—. *The Genealogy of Morals*. Translated and edited by Walter Kaufmann, Vintage Books, 1967.

—. *Thus Spake Zarathustra*. Translated by Thomas Common, Prometheus Books, 1993.

Pippin, Robert. *Hegel's Practical Philosophy: Rational Agency as Ethical Life*. Cambridge UP, 2008.

Raschid, Salman. *Iqbal's Concept of God*. Oxford UP, 2009.

Renan. Ernest. "What is a Nation?" Translated by Ethan Rundell. Presses-Pocket, 1992. Paris. http://ucparis.fr/files/9313/6549/9943/What_is_a_Nation.pdf. Accessed 28 Jan. 2018.

Said, Edward. *Beginnings: Intention and Method*. Columbia UP, 1985.

—. *Reflections on Exile and Other Essays*. Harvard UP, 2000.

Schneer, Jonathan. *London 1900: The Imperial Metropolis*. Yale UP, 1999.

Sevea, Iqbal. *The Political Philosophy of Muhammad Iqbal: Islam and Nationalism in Late Colonial India*. Cambridge UP, 2012.

Spengler, Oswald. *The Decline of the West*. Translated by Charles Francis Atkinson, Alfred A. Knopf, 1929.

Stone, Dan. *Breeding Superman: Nietzsche, Race and Eugenics in Edwardian and Interwar Britain*. Liverpool UP, 2002.

Trotsky, Leon. *Literature and Revolution*. Haymarket Books, 1925.

Vico, Giambattista. *New Science*. Penguin Books, 1999.

Vincent, Andrew and Plant, Raymond. *Philosophy, Politics, and Citizenship: The Life and Thought of the British Idealists*. Basil Blackwell, 1984.

Zakaria, Rafiq. *Iqbal: The Poet and the Politician*. Viking, 1993.

Žižek, Slavoj and Gunjevic, Boris. *God in Pain: Inversions of Apocalypse*. Seven Stories Press, 2012.

Part II
Literary history

5 Lu Xun's indigenous modernity

Philology and resistance in *Old Tales Retold*

Daniel Dooghan

China's comparative absence from postcolonial studies is a mark of failure. The failure is one of both vision and effort. That China was not colonized is false: those who have walked in the shade of Shanghai's beautiful plane trees did so in the formerly extraterritorial French concession. Perhaps it did not experience European colonialism to the extent of South Asia or Africa, but China was not immune to the ravages of empire. Paradoxically, despite the lack of broad focus on China in postcolonial studies, in a narrow sense China has been central to debates in the field through Fredric Jameson's deployment of the writer Lu Xun (魯迅) (1881–1936) in his controversial essay on third-world literature. In it Lu Xun emerges as the paradigmatic figure of a prominent configuration of world literature. For all the furor over the essay, Lu Xun did write national allegories, and European modernism remains an aesthetic standard for contemporary world literature. Tellingly, much critical effort has been expended to move Lu Xun from national allegorist to modernist (Tang 1232). This push to assert Lu Xun's aesthetic worldliness, paralleled by the coalescence of a postcolonial canon and the more nebulous inflection point between postcolonial and world literatures, reflects a desire to render these texts—the world—commensurable with metropolitan aesthetic practices. To be sure, these practices are not solely the possession of the metropole, emerging as they do from the dialectic of vernaculars encountering broadly privileged forms. The functioning of this dialectic was of great interest to Lu Xun and, as I will argue, to Left-philology; however, plumbing this dialectic requires time and discipline.

By contrast, voguish critical rubrics such as exoticism, trauma, or hybridity, though often ripe for parody from below, offer a royal road to worldliness for the metropolitan reader. Theorizing about texts of the periphery in this way makes them accessible through scholarly idiom, and often more crudely through commercial categories, but deracinates them and risks privileging criticism over context. In the absence of that context, as perhaps in an anthology, reading the worldly, modernist Lu Xun obscures his role as a local, anticolonial figure. Even the presentation of Lu Xun as national allegorist defangs his radical potential. The compatibility of this pedagogy with neoliberal politics might recall the complicity of the humanities with the projects of empire, particularly given the centrality of reifying and affirming the values of Euro-American hegemony, which as Giovanni

Arrighi points out is both a historical exception and under siege by an ascendant East Asia (42). World literature, if more broadly inclusive than the postcolonial, here appears to have abandoned the latter's contesting of history in favour of an accounting for the status quo.

Thus the failure of vision; one of effort may be to blame. Despite the global reach of empire, the postcolonial as regularly studied is only marginally polyglot, and world literature conspicuously less so. The Anglophone and Francophone are well represented, and though the Latin American Boom may have passed, Spanish holds its own. The colonized using the languages of the colonizers accounts for this in part, but this is not a universal phenomenon. Saadat Hasan Manto (Urdu) is unknown in comparison with Salman Rushdie.[1] Even Nobel Prize winner Rabindranath Tagore (Bengali) cannot contend with the Booker-backed juggernaut.[2] Moreover, Rushdie the winner of multiple international prizes likely cuts a different figure in world literature from Rushdie the chronicler of the Emergency. Language and history are markers of locality that resist subsumption by universalizing models of literature and require long apprenticeships to overcome. Given the bilingualism of colonial India's indigenous elite, we have a variety of accessible anticolonial works either originally in English or readily available in translation.

The same cannot be said for China due to the differences in its semi-colonial situation (Shih 36). Few texts from China appear originally in English, and those that do[3] often aim to exploit the aesthetic preferences of the world literary market: a metropolitan readership empowered by economic hegemony. The perceived difficulty of Chinese—overstated, likely for Othering effect—places translation in the hands of specialists and renders context largely inaccessible for casual or student readers. Lu Xun's assimilation into world literature as national allegorist and then as modernist makes him accessible to worldly readers as a Chinese writer, but one who circulates in a transnational economy of similarly nationalistic—in the first case—or introspective—in the second—writers. His most famous works, 'Diary of a Madman' (狂人日記) and 'The True Story of Ah Q' (阿Q正傳), are well suited to this task, as their employment of history and language does not obscure the narrative; these stories are widely anthologized. His pioneering collection of prose poems, *Wild Grass* (野草), has entered into the critical discourse on account of its opacity, fitting in with discussions of linguistic play in avant-garde modernisms (e.g. Kaldis 10–11).

These cosmopolitan readings of Lu Xun are well supported by his writings, or at least those in common translation and circulation. Lu Xun's corpus—even excluding his many translations—is vast, spanning eighteen volumes in the most recent Chinese edition. The few texts on which his international fame rests are a fraction of those that made him a literary celebrity during his lifetime. A prolific and aggressive essayist, as well as a ground-breaking scholar of Chinese fiction, Lu Xun is much more than a synecdoche for a variably modern Chinese literature. Although pedagogically convenient, the logic of rupture that so often attends to Lu Xun fails to capture the radicalism of his scholarship, essays, and translations that constitute the bulk of his output. These works highlight the dialectic between local and global in Lu Xun's thought. His translation practice served to facilitate global literary exchange, whereas his scholarship recast China's literary history, and his essays prosecuted his increasingly, if unconventionally, leftist beliefs amid local fights over the correct cultural line. What Lu Xun's corpus offers *in toto* is not an easily assimilable nationalism or modernism, but an erudite meditation on the role of the Chinese language and its literatures in a globalizing world. Through both a celebration of the vernacular in his fiction and scholarship, and a challenge to it in his translation, Lu Xun attempted to recast Chinese literary history in the hopes of changing its future. This is a local story that does not easily fit prevailing categories of world literature.

In the preface to his first collection of fiction, Lu Xun articulates his ambivalence over China's potential for reform. Often cited as evidence of his nationalism—correctly if all too narrowly—the preface narrates the author's youthful desire for national salvation first through medicine, then through literature: neither works. Still, the older, chastened Lu Xun concedes, "however hard I tried, I couldn't quite obliterate my own sense of hope" (*Real Story* 19). Failure may have dogged Lu Xun's efforts at revolutionary cultural change, like those of many contemporary idealists and intellectuals, but this reticent hope sustained a much larger project than inaugurating a new China with a new literature. The old China would not go away so easily; indeed, it is the primary subject of much of Lu Xun's fiction. Nor should it, since a blunt repudiation of tradition invites conservatives to defend their commitments. Instead, Lu Xun used his comprehensive knowledge of local literary history and his voracious reading of foreign literature to rewrite the conditions of possibility for Chinese literature and, hopefully, China itself. By focusing on the marginal and the vernacular, he presented an alternative literary history for China that admitted the legitimacy of the popular and established a local ground for his advocacy of a globally oriented leftism.

Inconveniently for this thesis, Lu Xun was not a systematic thinker. This is not to say that his formation cannot be mapped or that his beliefs lacked any consistency. However, the occasional nature of much of his writing—his essay genre, *zawen* (雜文), means miscellaneous writing—and its contingency on debates among Chinese intellectuals makes for much invective and not always for the sober and rational articulation of positions. While this productively underscores the role of the geographically local in discussions of world literature, it also complicates tracing a clean line of argument through his corpus. However, his last collection of stories, *Old Tales Retold* (故事新编), along with his—actually long—*Brief History of Chinese Fiction* (中國小說史略), offer sustained and understudied examples of his rewriting of Chinese literary history. The latter's absence from critical discourse is understandable given that it is not literature as conventionally understood. Nevertheless, the novelty of Lu Xun's project and its literal rewriting of Chinese literary history against the generic hegemony of poetry and history merit far more discussion than I fear can be offered here. In the *Old Tales*, history and language are inseparable from the narratives as they are densely allusive reworkings of Chinese myths. As with Dante, the material is opaque without either a command of the source material or an extensive apparatus. Although it contains many of Lu Xun's most complicated treatments of indigenous and foreign materials—it is also great fun—it languishes unstudied and unread as world literature in favour of texts more easily subsumed by the world literary market. The requisite knowledge of local literary history to parse the *Old Tales'* constituent narratives goes far beyond glossing the occasional allusion. I do not doubt that a canny reader could produce a reading of the *Old Tales* that spoke to a current critical fetish, but this has yet to occur.

Philologists for the future

The world literary market is obviously not the totality of literature, but as Francesca Orsini demonstrates, it shapes what texts are available to metropolitan subjects (327). Reading the local is difficult, but reading the world becomes easy because, in the literary field as in many others, it tends to revolve around those subjects. Metropolitan readers of world literature can see their interests and concerns reflected with flattering consistency in disparately sourced texts. But the metropole is not the world, and conflating the two constitutes an apology for cultural imperialism and worse. Assuming a metropolitan standard for international development, for example, has not proven especially beneficial for those being developed (e.g. Prashad 226). More texts from more places, while admirable, will not help if they are all mobilized to speak the same way. The prevalence of translations into imperial languages implicitly flatters metropolitan subjects: a reminder that in many practical ways the world revolves around them, at the expense of recognizing the contingency of their position.

As a well-intentioned metropolitan subject, the trick, then, is to decentre oneself. Given the multiple overlapping privileges of the metropolitan reader, this is probably impossible. Nonetheless, Timothy Brennan argues, we must try to correct a failure of vision through the application of effort. Methodology, for Brennan, requires looking differently; theory etymologically is, after all, a way of seeing. Brennan's goal is to "draw attention to a position and a politics developed in peripheral zones whose inspiration is as much European as non-European" (3). As both a reader and an object of reading, Lu Xun offers such a position and politics: he and his texts illustrate the negotiation of Chinese literary tradition, contemporary political exigencies, and transnational aesthetic discourses. Lu Xun is not an apologist for a triumphal (or traumatic) hybridity, with globalization as a *fait accompli*. He instead offers a critical position that does not assume the superiority of the cosmopolitan over the local, a presentist rupture over historical continuity. By foregrounding such tensions, Lu Xun holds out hope that the world could be otherwise.

Just as on the aesthetic front the challenge here is to envision a world literature worthy of the name, so too politically must we acknowledge the existence of the vernacular and the multipolar against the totalizing logic of a cosmopolitan modernity. Modernity, if we can speak of just one, is not simply arrived at, but produced continuously and with difficulty. Brennan reminds us, "Rather, modernity is singular because of the overdeveloped and interlocking global systems of capital, always the prime movers of colonialism and imperialism"; however, "a singular modernity, as the willing or unwilling outcome of capitalism, bears the stamp of its own diversity and resistance" (13). Uncovering this stamp requires a new—rather, an old—methodology. Articulating an intellectual lineage of erudite dissidence deriving from Giambattista Vico, Brennan calls for an innervated philology:

> The emphases of this hermeneutic may be said to be on the vulgate rather than the classical; on secular and corporeal solidarities rather than sacred textual encounters; and on the circulation of demotic and experimental forms rather than their containment within notions of aesthetic autonomy.
>
> (4)

The aesthetically autonomous, Theodor Adorno's privileged category, arrogates for an international high modernism, that which Brennan contends has more modest origins. Adorno's desire to insulate art from commercialism or fascism is understandable, but this comes at the expense of divorcing what may be popular, regional, or otherwise marginal forms from their communities of origin. In Adorno's favoured medium, how Hungarian is Bartók? Brennan's corrective is a "civic hermeneutics—an aesthetic and a style that conform more closely to the actual modes of non-Western or postcolonial literatures and the arts than do prevailing forms of European or American modernism" (4). I would contend that similar classicizing tendencies are present in non-Western and postcolonial literatures, but may be similarly contested. Lu Xun's own work pursues the demotic both contemporarily and historically in Chinese as well as in his broader reading. Local and global are relative terms.

But a philological engagement with the local, as Brennan argues, is out of fashion in world literature. Contemporary theories of world literature such as circulation, planetarity, translation, distant reading, and world-systems privilege texts that travel or appear to resist attribution of geographical or linguistic origin.[4] Rebecca Walkowitz's excellent, provocative *Born Translated* is a current example of the theoretical push against siting texts locally: "Above all, world-shaped novels explore translation by asking how people, objects, ideas, and even aesthetic styles move across territories, and how that movement alters the meaning and form of collectivity" (123). Walkowitz is no doubt correct here, but her presentist emphasis on the "world-shaped" novel delimits a mode of global literary circulation as distinct from local styles. The kind of intra-textual pluralism highlighted here, though, is hardly new.[5] The global Anglophone novel is not the totality of Anglophone literary production, and indeed may be a small if critically visible fraction. This disproportionate visibility, though, risks portraying the economic and political networks that enable the cosmopolitan Anglophone as necessary rather than contingent on historical conflicts. Nevertheless, Walkowitz offers a methodological imperative on this basis, "insisting that geopolitical conflicts, both small and large, need to be understood locally as well as comparatively" (126). Although a long tradition of leftist criticism has asserted the same, and novelty of the global Anglophone is suspect here, Walkowitz importantly brings this lens to mainstream studies of world literature.

This dialectical tension between the local and the global is what philology can recover. In articulating the plan for his *New Science*, Vico from the outset stresses the role of vernacular language in shaping the lives of local communities: "The mental vocabulary of human social institutions, which are the same substance as felt by all nations but are diversely expressed in language according to their diverse modifications, is exhibited to be such as we conceived it" (106). This contrasts with classicizing impulses to organize and dominate: the tyranny of the paradigm. The global Anglophone might be considered a classical English, which enjoys standardization and privileged reproduction combined with economic supremacy. What can be lost in the prestige of classicism is that its status is contingent on its development, rather than a teleological necessity. Following Vico, Brennan argues, "The stages of political hierarchy and counterhierarchy imprint their character on linguistic forms, and concealed political interests communicate themselves as modes of expression, quite apart from the content of the message" (41). For Vico, philology gives us the tools to recover the contingency of the classical from the ruins of the vernacular: "The great fragments of antiquity, hitherto useless to science because they lay begrimed, broken, and scattered, shed great light when cleaned, pieced together, and restored" (106). This is what Brennan takes as the counterhegemonic power of philology:

> The whole edifice of philological study oriented toward literary languages or to conferring status on ingenious individuals such as Sophocles, Vergil, or Catullus is subordinated here to the inventive exchanges of communities in the act of solving problems of development and organization, where active choice (language 'agreed upon') determines outcomes.
>
> (41)

Although Lu Xun is an archcanonical figure for multiple configurations of world—and Chinese—literature, his activity in such "inventive exchanges" is illustrative of Brennan's counterhegemonic philology, whose recovery poses a similar challenge to a centripetal world literature.

Lu Xun's late work offers an opportunity not only to implement this sort of critical philology, but also to see it at work historically. Granted, Lu Xun may not be the model philological leftist, *à la* Gramsci, or the model scholarly philologist, *à la* Brandes or Auerbach. For Brennan, however, Lu Xun could constitute a model philological figure, negotiating Chinese literary history, the newfound availability of foreign texts, and China's precarious geopolitical situation through his scholarship and creative work. Above all, his research on and production of vernacular fiction establishes Lu Xun as a practitioner of Brennan's "civic hermeneutics." However, this negotiation often gets swept up as a heroic origin of modern Chinese literature, with critics and propagandists both viewing it as an epochal rupture that effaces the human agency in play and packages it neatly for mythological deployment. At the Yan'an Forum (延安文艺座谈会) in 1942, Mao Zedong (毛澤東) took advantage of the latter to devastating effect, setting the agenda for the PRC's early aesthetic policy (Mao 483–484). As for the former, Lu Xun was all too human, with many of his works owing to his repressed elitism and cantankerous nature given voice by his celebrity. His concerns were human too, though. From his early plea to "Save the children" in 'Diary of a Madman' to his critical retellings of myth in *Old Tales Retold*, he recognized language—especially the vernacular—and literature as institutional forces in human lives, and endeavoured to intervene in them as he could (Lu Xun, *Real Story* 31). For example, Lu Xun began using the recently developed third-person feminine pronoun 她 in his writing and translations in 1924, a year after his major speech advocating for women's rights through a critique of Henrik Ibsen's *A Doll's House* (Gamsa 145). However, Lu Xun was not programmatic in his work; though his translations exhibit some systematic tendencies, he was an idiosyncratic and variable thinker. Still, his long apprenticeships in classical Chinese literature, Japanese, German, and other literatures in translation positioned him well to respond to the geopolitical situation on which his own learning was contingent.

That learning was extensive: Lu Xun knew philology. Still, he was not a professional like contemporaries Fu Sinian (傅斯年), a founder of Academia Sinica, or Gu Jiegang (顧頡剛), a major textual historian (Chang 327–328). He knew both men, though, as he was a literature professor who moved in scholarly circles. Lu Xun corresponded with Fu as early as 1919, and met him repeatedly in 1927, one year before the latter set up the philology institute at Academia Sinica (*Lu Xun quanji* 15:365; 16:7–8, 11–12, 16). Gu Jiegang receives a fair bit of criticism in the *Old Tales*: he appears as a character named 'Birdbrain' as a malicious play on his historiographical methods (*Real Story* 320; Anderson 259n1). This is not to say that Lu Xun rejected linguistic or textual mastery, but that he did not see it as an end in itself.

He was no slouch, though. His early training in classical Chinese texts ensured at least a working familiarity with textual scholarship, because those texts are unreadable without resort to an intra-textual apparatus. Moreover, he was a pioneering scholar of Chinese fiction. In his *Brief History of Chinese Fiction*—the first work of its kind—he makes use of the comparative method (179–182), weighs in on the study of Chinese phonology (311–312), engages in Vichian speculation about the origins of fiction (275), and indicates his keeping up on scholarship as late as 1935 (422). Less well known than his research into literary history are his editions of ancient Chinese fiction. Throughout his life he corrected and edited many classical texts, and wrote many paratexts for them (*Lu Xun quanji* 10:1–160). His lecture notes on Chinese literature from his brief tenure as a professor at Xiamen University in 1926 cover early texts through Sima Xiangru (*Lu Xun quanji* 9:351–442). Again, while he lacked the training in contemporary linguistic philology that Fu Sinian gained in Europe, his literary erudition was vast and conversant with philological methods (Chang 313–316). Similar to Vico's, Lu Xun's philology inaugurates the study of Chinese fiction as an institution.

Looking at fiction as a human institution enabled Lu Xun to present a neglected textual tradition—*xiaoshuo* (小說), literally little talk—as an alternative history. This is in keeping with the indigenous philological scene that developed during the late Qing: the *kaozheng* (考證), or seeking evidence school. Benjamin Elman links this strand of scholarly scepticism about antiquity to the "cultural icono-clasm" of Lu Xun's era, but cautions that those researchers "remained committed to classicist ideals" rather than a proleptic radicalism (Elman 244). This parallels the awkward situation of Lu Xun's generation, which "was transitional, and knew itself to be so" (Schwarcz 28). Much of Lu Xun's early fiction features characters caught between tradition and an ungrasped modernity, turning to an ungraspable modernity by his middle period. As the enthusiasm of the May Fourth Movement waned, Lu Xun found tradition harder to escape. In a story from this period, 'Upstairs in the Tavern', a Lu Xun-type narrator listens to an old acquaintance lament the futility of trying to change against the weight of tradition:

> "You're teaching the classics?" I asked in surprise. "Of course, what did you think—that I was teaching English? I started off with two students, one doing *The Book of Odes* and the other *Mencius*. I've just got a new one, a girl— she's studying *Classical Maxims for Young Ladies*. I don't even teach maths: not because I don't want to, but because they don't want it." "I never thought you'd end up teaching this sort of stuff." "It's their father who wants it—I'm just the hired help, I don't care. What a waste of time it all is. But I get by."
>
> (*Real Story* 187)

Although classical texts are singled out here as indicative of tradition's inertia, they represent only one possible Chinese tradition: imperial orthodoxy. This period coincides with Lu Xun's research into traditional fiction, which, as he would demonstrate in the *Old Tales*, can offer a different narrative of Chinese history. Like the earlier *kaozheng* school, he sought to ground Chinese literary history on facts rather than accreted interpretations, enabling tradition to be a ground for change rather than its limit. Unlike critical narratives of heroic nationalism or aesthetic modernism, Lu Xun's lifelong engagement with Chinese literary history rests on a logic of continuity rather than rupture.

Peripheral imperialisms

Although rupture is a compelling logic in nationalist interpretations of Lu Xun, preferring instead a causal continuity reveals his potential as a thinker of the local–global dialectic and anticolonialism. As a subject of the Manchu Qing dynasty, the man who would become China's most prominent modern writer was born Zhou Zhangshou (周樟壽) in 1881 amidst peripheral imperialisms then active in northeast and central Asia (Pollard 16–17).[6] China had recently consoli-dated its hold over Xinjiang[7] through the 1877 victories of General Zuo Zongtang (左宗棠),[8] but had lost only two decades earlier its claim over what became part of the Russian Far East. While acting regionally as both imperial aggressor and victim, China in its internal politics evinced another form of imperial domination. Since 1644, erstwhile foreign Manchurians had ruled the Chinese empire, having exploited the chaos at the end of the Ming dynasty (1368–1644) to seize power. The Manchus imposed extensive social and cultural policies that restricted the expression of Han Chinese identity.[9] From birth Lu Xun was enmeshed in the imperialist politics that would shape his creative life.

How he came to that life is well known thanks to the autobiographical preface of his first collection of stories. 1923's *Nahan* recounts his family's travails and his nascent interest in Western science. Yet in looking to science, the scion of the fallen Zhou family looked east towards Japan. He saw China's ancient medical sciences as fraudulent and possibly culpable in his father's death, whereas Japan was undergoing a renaissance in breaking from its traditions: "The translated histories I read, meanwhile, informed me that much of the dynamism of the Meiji Restoration sprang from the introduction of Western medicine to Japan" (Lu Xun, *Real Story* 16). His father's death in 1896 came on the heels of China's defeat in the Sino-Japanese War. Merely twenty-seven years after the Meiji Restoration and just over four decades after Commodore Perry's arrival in Japan forced an end to the isolationism of the Tokugawa period (1603–1868), Japan had become a player in imperial geopolitics. Confirming its imperialist debut, Japan, like Britain before it, attacked and defeated China.

This marked a staggering reversal of fortune for China, which had been the cultural hegemon of East Asia for millennia. Japan extracted Korea from the Chinese tributary system, paving the way for its eventual annexation in 1910. Taiwan also went to Japan in the Treaty of Shimonoseki that ended the war in 1895, leading tortuously to the present conflict over the island's political status. Of course, China had also been an imperialist power since the Han dynasty (206 bce to 220 ce), but Japan's victory signified more than a changing of the guard in East Asia. China's tributary empire dated from antiquity and was qualitatively different in its conceptualization of the world from Euro-American imperialisms of the nineteenth century (Chin 12–15). Japan, by contrast, employed a sophisticated technological and ideological apparatus to dominate its colonies. For example, like other contemporary imperial powers, Japan used photography in Taiwan to promote its cultural policies (Allen 1013). China had not only been defeated militarily, but was also out of date.

In response, the Qing government made some abortive attempts at change. The Hundred Days' Reform of 1898, led by Kang Youwei (康有為) and his student Liang Qichao (梁啓超) under the aegis of the Guangxu Emperor (光緒帝), was crushed by the Empress Dowager Cixi (慈禧太后). Kang and Liang—both philologically minded scholars—would continue as major figures in China's modernization, though without the support of the Qing. Despite its opposition to institutional reform, the Qing did fund scholarships for students to study abroad (Deng 38). As with the later Boxer Indemnity Scholarship Program, many of the recipients of these scholarships became leading figures of the Republican Era (1912–1949). Among them was Lu Xun.

First leaving China in 1902, he would spend seven years in Japan, but only studied for about half that time. Against the backdrop of imperial humiliation that enabled his studies, Lu Xun, perhaps unsurprisingly, became caught up in the revolutionary ferment brewing among his fellow expatriate students. As Japan triumphed in the Russo-Japanese War, the comparative weakness of China became even more glaring to these nationalistically minded students. Reports of the war prompted Lu Xun's Damascene moment that led him to abandon his medical studies for literature (Lu Xun, *Real Story* 17). Yu Dafu's (郁達夫) canonical 'Sinking' (沈淪) gives a more psychologically detailed if pathologically Wertheresque account of student life in Japan, focusing on feelings of impotence. Lydia Liu has further shown the deleterious influence of racist "national character" discourses on Chinese intellectuals during this period (45–76). Like other students, Lu Xun saw Japan not as a colonialist oppressor but as an example of modernity and turned his anticolonial ire towards the Qing government as the cause of China's troubles.

The Qing monarchs were indeed foreign, and their rule constituted a form of imperial domination that has recurred throughout Chinese history. Invading Khitans, Jurchens, Mongols, and here Manchus have all ruled a Han majority. However, these foreign conquerors tended to adapt to the Chinese rather than the other way around. Exceptions exist, such as the Manchus' imposition of the queue, but the Sinicization process tended to bring the rulers closer to the ruled in contradistinction to most imperial projects. Thus the turn against the Qing government was relatively sudden and perhaps exacerbated by the painful presence of so many other signifiers of China's now-multinational imperial domination.

The queue became a chief signifier of Manchu domination. Although, as Eva Shan Chou points out, the queue had enjoyed centuries of assimilation and even cultivation, its "normalcy" faded as its utility as a symbol of anticolonial resistance grew (Chou *Memory* 79–80). Lu Xun could not bring himself to violent struggle, but he, like many others, could cut his queue (Chou, *Memory* 74). The queue would go on to figure in Lu Xun's longest story, 'The True Story of Ah Q', as an ironically empty symbol of political affiliation during the Xinhai Revolution (辛亥革命) that overthrew the Qing in 1912. His youthful enthusiasm for intellectual revolution, itself inspired by a transnational romanticism, expressed in 1908's 'On the Power of Mara Poetry' (魔羅詩力說), faded as the cutting of the queues and revolutionary politics did not solve China's underlying internal issues.

Much of Lu Xun's work would critique the efficacy of the Xinhai Revolution, because the shift in government did not correspond to China's liberation from foreign domination. Ousting the Manchus did not restore Taiwan to China, end the unequal treaties, or win any geopolitical clout. At the Versailles Conference in 1919, Germany's possessions in China did not retrocede to China but were instead awarded to Japan, leading to student protests on May 4 of that year. This event became the catalyst for a range of literary activity in China. Lu Xun began writing vernacular fiction on critical themes in 1918, though his readership and celebrity did not mature until the May Fourth Movement was in full swing (Chou, "Learning" 1043). By 1926, though, even this enthusiasm had waned. For all the fervour among the literati, change was slow to come. That year Lu Xun prefaced *Panghuang* (彷徨), his second collection of stories, with a quotation from Qu Yuan's (屈原) 'Li sao' (離騷), an ancient poem lamenting the failures of government and their toll on the poet (Lu Xun, *Real Story* 169).

That Qu Yuan committed suicide would not have been lost on Lu Xun's readers. Apart from a slim volume of prose poems the following year, Lu Xun's creative life seemed to have ended. Those poems continue the dark themes presaged in *Panghuang*.

Conflicts over translation

Although he stopped writing fiction, jaded to the possibility of the literary revolution he envisioned as a youth, Lu Xun continued translating and writing polemical essays in response to local aesthetic and political debates. In selecting foreign texts for translation and by promoting foreign aesthetic principles he showed both an engagement with world literary discourse and a critique of it: his selections were not neutral. Yet Lu Xun was only one among many in a fractious group of literary intellectuals during the Republican Era, and his stature made him a target for opposing factions. His idiosyncratic leftism—a vague commitment to proletarian literature brought about in part by his translations of Anatoly Lunacharsky—also brought him into conflict with both more orthodox Marxists and conservatives. Pugnacious as ever, though, Lu Xun always fought back. He would drift into occasional coalitions with the broader Left, through the League of Left Wing Writers for example, but had harsh words for those who espoused what he saw as a reactionary aesthetic line. In this latter category he placed the members of the Crescent Moon Society, a faction advocating for a literature of universal appeal against the political focus of the leftists.

His enthusiasm for Soviet literature and aesthetic theory, pursued through his translation programme, indicates that he had not entirely abandoned his earlier interest in revolution. However, his attacks on the Crescent Moon Society explicitly link his later literary activities with an anticolonial struggle that has unexpected resonances with debates about world literature today. He saw the Crescent Moon members as constituting a world literature—largely Anglophone—that flattered their view of the world.

Lu Xun traded barbs in periodicals with Crescent Moon affiliate Liang Shiqiu (梁實秋) over the political uses of literature. Liang, Lu Xun argues, advocated for a literature of "human nature" based on "common features" among classes against the latter's class-based aesthetic theory (*Selected Works* 3:85). He casts Liang's concern for a shared humanity as ersatz, seeing it instead as cover for a defence of bourgeois values (*Selected Works* 3:85). This analysis hinges on Liang's rejection of literature as a tool of class struggle and a mercenary conflation of a Marxist species-being with bourgeois comfort:

> As a case in point take this article by Mr. Liang, aimed at doing away with the class character of literature and blazing the truth abroad. We can see at a glance that this view of property as the base of culture and of the poor as scum doomed to extinction is the 'weapon'—I mean the 'reasoning'—of the bourgeoisie.
>
> (*Selected Works* 3:88)

Gloria Davies sees Lu Xun as going so far as to paint the Crescent Moon group as in league with imperial domination: "He sneered that, in advertising themselves as sole agents for their chosen foreign luminaries, the Crescent Moon members resembled the 'cordon of interpreters, detectives, police "boys" and so on' who served their European masters in Shanghai's foreign concessions" (108). A world literature that trades in discussions of universal human rights without any critique of why those rights may not actually be universal, then, is for Lu Xun little more than a self-congratulatory apologetics for a privileged class. Despite having limited formal engagement with Hegelian and Marxist dialectics, through his leftist sympathies and vernacular commitments he reaches a similar anticolonial position with regard to the constitution of world literatures.

Hyperbole of the polemic aside, what appears to irk Lu Xun the most here is the lack of self-awareness on the part of the Crescent Moon group. Whether or not they were really the running dogs that Lu Xun insinuates is difficult to assess given its heterogeneous membership.[10] That the occasion for the attack was a debate over methodology in translation points to Lu Xun's belief in both the necessity and possibility of cultural change through careful introspection. His philosophy of "hard translation,"[11] a method of radical fidelity to the source language, was grounded in a desire for linguistic change brought about by foregrounding the difference between the source language and Chinese, hinting at the inadequacy of the latter:

> And now that we are dealing with 'foreign languages' we may need many new forms of construction—which, to put it strongly, have to be made by 'hard translation'. In my experience, you can retain the flavor of the original better by this method than by rearranging your sentences; but modern Chinese has its limitations because it is still waiting for new constructions. There is nothing 'miraculous' about this. Of course, for some people it is no 'fun' to have to 'trace with a finger' or 'make a mental effort'. But I had no intention of giving these gentlemen 'pleasure' or 'fun'.
>
> (*Selected Works* 3:81)

Reading in translation, for Lu Xun, should be laborious because it is simultaneously a reflective and pedagogical activity. The easy apprehension of texts is inimical to his goals because the pursuit of such readerly pleasure allows the reader to avoid any self-criticism on the linguistic level. Given the politically charged aesthetic theory Lu Xun was translating, such as that of Lunacharsky, and his political aims, in his eyes easily consumed translations likely enabled readers to gloss over challenging content as well (*Selected Works* 3:75). Lu Xun's view of world literature is predicated on decentring the reading subject; doing otherwise reinscribes systems of domination, internal and external.

Lu Xun's "hard translation" provided one method of negotiating foreign influence with an eye towards reshaping the domestic. His resistance to imperial domination was tempered by his recognition of the reality of the globalizing world. He spent the last years of his life on the edge of the foreign concessions in Shanghai, to which he calls attention through the shared title of three late essay collections: *Qiejieting* (且介亭). This is a play on *zujie* (租界), or foreign concession. By writing half of each character he creates a visual pun loosely translating to 'semi-concession studio' (*Lu Xun quanji* 6:5n9). This indicates the contingency of his career on the foreign—he got his start in imperial Japan, and the Shanghai concessions provided some sanctuary when he offended the Chinese authorities—while highlighting its inaccessibility. That translation work dominated his productive life also indicates his awareness of the need for interfaces between the local and the foreign. His chosen method maintains their distance, in opposition to those who would uncritically align themselves with a tempting image of modernity. Lu Xun saw such easy moves as representing illusory change. Instead he chose the long route of working through language to show the contingency of cultural institutions at home and abroad.

In reading Lu Xun as a model *for* reading world literature, and as a model *of* reading world literature, we would do well to follow his example and take up the challenge of philology. To this end, let us turn to how Lu Xun's *Old Tales Retold* deployed his overwhelming literary erudition both to respond to China's increasingly, if aberrantly, peripheral status, and to critique the responses of others. He accomplishes this not by producing a self-consciously cosmopolitan literature, but by rehabilitating the Chinese classical tradition he had previously attacked. The stories in the *Old Tales* do not intimate a local atavism or assume a Platonically worldly audience, but weave together texts past and present, local and foreign to produce a critique of modernity contingent on the Chinese situation. As much as modernity is a global phenomenon, understood in its totalizing economic and political logics, it is not simply given; the *Old Tales* reminds us that local histories and global futures are not incompatible, and that the narrative of their dialectic can be contested.

Old Tales Retold, or how one philologizes with a hammer

At the end of his life, Lu Xun turned to mythology. The collection he produced, *Old Tales Retold*, is an anomaly in his corpus. Its stories are well known in the Chinese mythological tradition, but Lu Xun's rewritings cast them in novel contexts and with deep erudition. Of the eight stories, five are from 1934 or 1935, leaving two from 1926 and one from 1922. This last piece was originally part of his first story collection, but cut from later editions. He plays with mythological revisionism a couple more times before setting to it in earnest in late 1935, suggesting that this was a deliberate project that took time to germinate. His preface concludes with a curious self-debasement:

> Most of the pieces are only sketches, and certainly not literary fiction. At times I base myself in historical fact; at others, my imagination roams free. And because I can't convince myself that the ancients are as worthy of respect as my contemporaries, I've found myself periodically slipping into the quicksands of facetiousness.

> (*Real Story* 296)

Julia Lovell follows convention in rendering *youhua* (油滑) as 'facetiousness', though earlier links it with 'slippery', showcasing its resonances with oiliness (*Real Story* 295).[12]

This slipperiness is central to the *Old Tales*, as indicated by Lu Xun's repeated invocation of it in the short preface. Similarly, Lovell's "imagination roams free" captures the playful Zhuangist[13] echoes of the original's *xinkou kaihe* (信口開河), connoting spontaneity and irresponsibility in speech (*Real Story* 296; *Lu Xun quanji*, 2:342).[14] Despite this professed playfulness, Lu Xun's sources are rigidly canonical, and, as Wilt Idema shows, backed up by extensive textual knowledge: Lu Xun draws on the multiplicity of retellings in Chinese literary history (Idema 59n121). Yet, by stripping his stories of their imperial interpretations and making them speak critically of authority, these tales offer what Eileen Cheng calls an example of Mikhail Bakhtin's carnivalesque (182). They illustrate the contingency of textual authority while not entirely abandoning it. Cheng links this to his transitional position in Chinese history, arguing, "In a sense, then, rewriting old tales may have given Lu Xun a 'viable space' that enabled him to reconnect with his deep-seated ambivalence about the past, even as his political stance was decidedly oriented toward the future" (170). This is criticism of tradition, to be sure, but not a dismissal. Rather than the direct assault on tradition mounted by stories like 'Diary of a Madman',[15] the pieces in the *Old Tales* rely on tradition without bowing to it.

Lu Xun's slipperiness in the *Old Tales* is not limited to Chinese literary history. Although the collection displays its author's vast understanding of a local textual tradition in both its historical and contemporary manifestations, it also reveals an engagement with global literary production. Lu Xun comfortably folded diverse foreign material into the *Old Tales*; however, unlike the Crescent Moon Society, his inclusions illustrate the dialectic between the independently local and the imperially global rather than an alignment with and apologetics for the latter. Although Lu Xun's engagement with foreign literatures was extensive, two examples that bookend the collection stand out as especially salient. First, Sigmund Freud appears in the preface, cited as a motivation for writing the first story (*Real Story* 295). Although that story, 'Mending Heaven' (補天), does not strike me as especially psychoanalytical, his citation indicates a participation in a global, contemporary critical discourse.

Second, the collection's final story, 'Resurrecting the Dead' (起死), is written as a one-act play. Although China has a rich dramatic tradition, its conventions differ from the play as understood in European literature, which was then a recent arrival in China (Idema 256). Without comment, Lu Xun employs the imported form to structure content drawn from the ancient *Zhuangzi*. Like many intellectuals in China—and worldwide—Lu Xun embraced Ibsen, though this piece is not especially Ibsenesque (*Selected Works* 2:85–92). Instead, it uses a form that is available to participants in global literary exchange. The combination of foreign form and local content does not appear as problematic here, nor should it given its contemporaneity with Cao Yu's *Thunderstorm* (雷雨), a smashingly successful four-act play. The collection draws from foreign literatures without becoming imitative. The stories in the *Old Tales* are not retold as something else—perhaps like *Ulysses*—or in the style of someone else—for which he critiqued the Crescent Moon faction. Instead they are simply retold at a different historical moment, in which China and Chinese literature are participating in a more fully globalized, though not homogenized, literary discourse.

We might debate the success of the *Old Tales* because of its obscurity. Whatever Lu Xun's plan was for it, the book does not appear to have found its audience. This may actually be an indication of its methodological efficacy, but also of the difficulty in following that method. Lu Xun's philological efforts into Chinese fiction showed him the variations and multiplicities that existed among texts as well as the various uses to which those texts were put. His *Brief History* goes to great lengths to show its readers how texts stabilized over time rather than appearing as oracular pronouncements. *Old Tales Retold* introduces slipperiness into literary history by restoring instability to canonical works. The atomic, paratactical nature of many ancient Chinese texts would seem to belie their canonical stability. However, the institutional practices of literature glossed over—literally—these texts. Yet even these glosses were historically contingent: for example, Kong Yingda's Tang-era commentaries gave way six centuries later to those of Zhu Xi. Imperial examinations tested textual knowledge, true, but through the lens of the authorized interpretation. Even the rote learning of the classics militated against seeing them as anything but stable. Thanks to his deep erudition and historical moment, Lu Xun may have been uniquely equipped to see Chinese literary history as open to revision on the scale of the *Old Tales*. Rather than integrate China into a world literature, the *Old Tales* reveals the many worlds of Chinese literature and *a fortiori* the many worlds of world literature.

Conclusion

The *Old Tales* is less about rewriting history than showing its contingency. Lu Xun's long apprenticeship studying marginal, non-institutional forms of Chinese literature enabled his critique of Chinese institutions, and his experience of empire gave him a reason to make it. His indirect attack on tradition conserves it while opening it up to revision: China's long history did not have to be abandoned to achieve modernity. The *Old Tales* argues instead that China can respond to the globalizing world on its own terms. Lu Xun's modernity is Chinese at heart; it is not borrowed, appropriated, or imposed from without. Though Lu Xun's path to modernity may have been idiosyncratic, it illustrates the potential of humanistic scholarship to expose the contingency of popular discourses. Perhaps most importantly, it speaks to the power of human institutions—literature, language, and scholarship—to shape the lives of humans. Lu Xun's advocacy of proletarian literature, his roundabout anticolonialism, and his goals for his literary practice testify to his commitment to critiquing human institutions for humane ends. As the *Old Tales* show, world literature, translation, and local literary history thus all emerge as subjects for Lu Xun's critical philology. Had he not died in 1936, perhaps he could have further opened up this productive space between scholarly humanism and doctrinaire leftism.

Today, Timothy Brennan sees philology as a bulwark against the various anti-humanisms now prevalent in scholarly and popular discourses. The adoption of positions that dethrone the human as the subject of history, though potentially grounded in good intentions, trades a project of universal liberation for an apologetics of consumer identity under guises of ecological responsibility or technofuturism (Brennan 7). Getting at why misanthropic ideas persist is central to Brennan's project, but more importantly, he asks—after Marx—what else is possible: "which traditions open up possibilities for a transformed future" (14). Philology, with its focus on human institutions, is the means to accomplish this "by directing us back to the human past, which is the only one we have that we can make again" (14). The challenge of philology begins with the technical, but does not end there; it allows us to understand the contingency of our world, and thus the possibility that it could be different.

Lu Xun's mythological turn is this kind of philology, and serves as a reminder that world literature is such only if we recognize a plurality of possible worlds. World literature as a theory is necessarily abstract, but as such has the ability to gloss over much of what makes reading it interesting. Worse, the glosses tend to reflect the projects of the glossers: for world literature, this is often the metropolitan Anglosphere. At risk are the worlds of human complexity, predicated on local differences—axiomatic in Vico. Lu Xun's devotion to language and literature may not have been entirely fruitful, but a philologist's life has no guarantee of success. Still, as his example shows, it offers a method and a challenge to see the world in its radical contingency and thus to resist anti-humanistic claims of necessity and inevitability.[16]

Notes

1 Although hardly conclusive, Google Scholar lists 1,310 hits for Manto and 38,600 for Rushdie.
2 Tagore manages 26,800 hits on Google Scholar, though over a century compared to Rushdie's four decades.
3 Pearl Buck and Lin Yutang's mid-century works would fit; Zhang Yijia's memoir *Socialism Is Great!* might serve as a more contemporary example.
4 I derive these categories from landmark statements on world literature by David Damrosch, Pascale Casanova, Emily Apter, and Franco Moretti, among others. Recent work by Alexander Beecroft and Eric Hayot offers alternatives to a dichotomy between global and local texts.
5 *Gilgamesh*'s preface refers to lapis lazuli, which was mined in Afghanistan, not Mesopotamia. Tiresias' admonition to Odysseus (11.126–34) prefigures the Great Commission (Matt. 28:19).
6 Zhou Shuren (周樹人) is the 'real name' most associated with Lu Xun in both Chinese and non-Chinese contexts. However, literati naming practices are complex, and Zhou Zhangshou was his earliest name.
7 新疆, literally 'new frontier'.
8 From whom the popular Chinese-American chicken dish tenuously derives its name.

9 These were so successful that many Manchurian customs are seen worldwide as authentically Chinese. Best known among these is likely the 'Mandarin' gown, of which the modern *qipao* or cheongsam is a derivative.

10 Perhaps tellingly, Liang Shiqiu went with the Nationalists to Taiwan, as did Hu Shi: another leading figure of the Crescent Moon group. However, affiliate Shen Congwen remained in the PRC despite running afoul of the political line.

11 The 'hard' in hard translation, 硬譯, signifies physical rather than metaphorical hardness—though it is difficult as well. As the following passage demonstrates, the goal of this method was to produce a translation that presents a hard surface to the reader as an obstacle to facile readings.

12 油滑*youhua* is literally 'oil slide', which has a range of connotations similar to 'slippery' or 'greasy' in English. I thank Joe Allen for calling attention to this phrase, and for introducing me to the *Old Tales*.

13 Although the *Zhuangzi* does not appear frequently in Lu Xun's corpus—the *Old Tales* being a notable exception—the Daoist classic appears to have been a favourite. Given Lu Xun's temperament, this is not surprising.

14 The Chinese idiom is equivalent to 'running one's mouth'.

15 The critique in 'Madman' may be less forthright than stated here, given ambiguities in its frame narrative and Lu Xun's complex relationship with tradition; however, nearly a century of commentators have seen the iconoclasm in this work, even if it has recently become more complicated.

16 I would like to thank my research assistant Noah Oakley for his help in preparing the manuscript.

Works cited

Allen, Joseph R. "Picturing Gentlemen: Japanese Portrait Photography in Colonial Taiwan." *The Journal of Asian Studies*, vol. 73, no. 4, 2014, pp. 1009–1042.

Anderson, Marston. "Lu Xun's Facetious Muse: The Creative Imperative in Modern Chinese Fiction." *From May Fourth to June Fourth: Fiction and Film in Twentieth-Century China*, edited by David Der-wei Wang and Ellen Widmer, Harvard UP, 1993, pp. 249–268.

Arrighi, Giovanni. "The Rise of East Asia and the Withering Away of the Interstate System." *Marxism, Modernity, and Postcolonial Studies*, edited by Crystal Bartolovich and Neil Lazarus, Cambridge UP, 2002, pp. 21–42.

Brennan, Timothy. *Borrowed Light: Vico, Hegel, and the Colonies*. Stanford UP, 2014.

Chang, Ku-Ming Kevin. "Philology or Linguistics? Transcontinental Responses." *World Philology*, edited by Sheldon Pollock, Benjamin A. Elman, and Ku-ming Kevin Chang, Harvard UP, 2015, pp. 311–331.

Cheng, Eileen J. *Literary Remains: Death, Trauma, and Lu Xun's Refusal to Mourn*. U of Hawai'i P, 2013.

Chin, Tamara T. *Savage Exchange: Han Imperialism, Chinese Literary Style, and the Economic Imagination*. Harvard University Asia Center, 2014.

Chou, Eva Shan. "Learning to Read Lu Xun, 1918–1923: The Emergence of a Readership." *The China Quarterly*, no. 172, 2002, pp. 1042–1064.

—. *Memory, Violence, Queues: Lu Xun Interprets China*. Association for Asian Studies, 2012.

Davies, Gloria. *Lu Xun's Revolution: Writing in a Time of Violence*. Harvard UP, 2013.

Deng Shaohui. "Government Awards to Students Trained Abroad, 1971–1911." *Chinese Studies in History*, vol. 28, nos. 3–4, 1995, pp. 35–48.

Elman, Benjamin A. "Early Modern or Late Imperial? The Crisis of Classical Philology in Eighteenth-Century China." *World Philology*, edited by Sheldon Pollock, Benjamin A. Elman, and Ku-ming Kevin Chang, Harvard UP, 2015, pp, 225–244.

Gamsa, Mark. *The Chinese Translation of Russian Literature: Three Studies*. Brill, 2008.

Idema, Wilt L. *The Resurrected Skeleton: From Zhuangzi to Lu Xun*. Columbia UP, 2014.

Jameson, Fredric. "Third-World Literature in the Era of Multinational Capitalism." *Social Text*, no. 15, 1986, pp. 65–88.

Kaldis, Nicholas A. *The Chinese Prose Poem: A Study of Lu Xun's* Wild Grass (Yecao). Cambria, 2014.

Liu, Lydia. *Translingual Practice: Literature, National Culture, and Translated Modernity—China, 1900–1937*. Stanford UP, 1995.

Lu Hsun. *A Brief History of Chinese Fiction*. Translated by Yang Hsien-yi and Gladys Yang, Foreign Languages P, 1976.

Lu Xun. *Lu Xun Quanji*. Renmin Wenxue Chubanshe, 2005. 18 vols.

—. "On the Power of Mara Poetry." Translated by Shu-ying Tsau and Donald Holoch. *Modern Chinese Literary Thought: Writings on Literature, 1893–1945*, edited by Kirk A. Denton, Stanford UP, 1996, pp. 96–109.

—. *The Real Story of Ah Q and Other Tales of China*. Translated by Julia Lovell, Penguin, 2010.

—. *Selected Works*. Translated by Yang Xianyi and Gladys Yang, Foreign Languages P, 1980. 4 vols.

Mao Zedong. "Talks at the Yan'an Forum on Literature and Art." *Modern Chinese Literary Thought: Writings on Literature 1893–1945*, edited by Kirk A. Denton, Stanford UP, 1996, pp. 458–484.

Orsini, Francesca. "India in the Mirror of World Fiction." *Debating World Literature*, edited by Christopher Prendergast, Verso, 2004, pp. 319–333.

Pollard, David E. *The True Story of Lu Xun*. The Chinese UP, 2003.

Prashad, Vijay. *The Darker Nations: A People's History of the Third World*. The New P, 2007.

Schwarcz, Vera. *The Chinese Enlightenment: Intellectuals and the Legacy of the May Fourth Movement of 1919*. U of California P, 1986.

Shih, Shu-mei. *The Lure of the Modern: Writing Modernism in Semicolonial China, 1917–1937*. U of California P, 2001.

Tang Xiaobing. "Lu Xun's 'Diary of a Madman' and a Chinese Modernism." *PMLA*, vol. 107, no. 5, 1992, pp. 1222–1234.

Vico, Giambattista. *The New Science of Giambattista Vico*. Translated by Thomas Goddard Bergin and Max Harold Fisch, Cornell, 1984.

Walkowitz, Rebecca. *Born Translated: The Contemporary Novel in an Age of World Literature*. Columbia UP, 2015.

Yu Dafu. "Sinking." Translated by Joseph S.M. Lau and C.T. Hsia. *The Columbia Anthology of Modern Chinese Literature*, 2nd ed., edited by Joseph S.M. Lau and Howard Goldblatt, Columbia UP, 2007, pp. 31–55.

6 Circuits of influence

Brodsky's Platonov and the ontology of alienation

Djordje Popović

In keeping with Timothy Brennan's call to make visible the philosophical traditions informing theoretical, political, and literary affiliations in the twentieth century, this chapter tells the story of an 'encounter' between a communist author (Platonov), an émigré poet and literary critic (Brodsky), and a right-wing philosopher (Heidegger), and of violence to which 'writers from the other Europe' were subjected upon their admission into the modernist canon. This particular episode begins with the 1973 publication of Joseph Brodsky's introduction to the first American edition of Andrei Platonov's *The Foundation Pit*. Platonov's dialectical and deeply humanist literary intervention against the ascendancy of economic and scientific determinism on the left—an intervention not unlike those of Marx, Lenin, Bloch, Lukács, Benjamin, and many others—becomes, under Brodsky's supervision, a masterpiece of dystopian fiction and a cautionary tale of humanist folly. Helping Brodsky in this endeavour is the political ontology of Martin Heidegger, whose theoretical priorities we will slowly begin to recognize in the language and manner of Brodsky's argument.

Two Platonovs

It turns out that the most important prose writer of the revolutionary generation was indeed an engineer—a Soviet *elekrifikator* and *meliorator*, Andrey Platonov (1899–1951). Today he is best known for the unconventional works he was not allowed to publish in his lifetime and, in particular, for the three novels he wrote prior to the 1934 Congress of Soviet Writers: *Chevengur* (1926–1929), *The Foundation Pit* (1929–1931), and the unfinished *Happy Moscow* (c. 1933). Having made a 'conformist' turn to socialist realism in the mid-1930s, Platonov remained active in Soviet literary circles, taking part in writers' brigades to Turkmenistan in 1934 (with Maxim Gorky's assistance), working with Georg Lukács, Mikhail Lifshitz, and others at *Literaturnyi kritik* from 1936–1940, and serving as a war correspondent for *Red Star* (on Vasily Grossman's recommendation). The unpublished novella *Soul*, the 1937 collection *The River Potudan*, and his 1947 short story 'The Return' are some of his more 'conventional' works from this latter period.

By the early 1970s, *Chevengur* and *The Foundation Pit* began appearing in *tamizdat*, which is to say they were published abroad from manuscripts smuggled out of the Soviet Union where the two novels would not be available for another twenty years. A publication history that was already punctuated by Soviet literary politics thus became even more complicated. Two separate canons emerged and a myth of 'two Platonovs' was born: a modernist and a (socialist) realist Platonov, with a line of demarcation running through the transitional works where, depending on one's critical perspective, Platonov's poetics either succumbs to external pressure or matures in response to the exigencies of time and Party.[1] Unsurprisingly, the problem of situating Platonov's work in his own time came to dominate Platonov scholarship from its earliest emphases on his famously idiosyncratic language and his relationship to the socialist state and its ideal, to the later and more nuanced studies of his narrative technique and philosophical themes latent in his work.

The latest attempts to "break [Platonov] out from the Soviet context [and into] the arena of world literature" and thus, as we shall see, to fulfil the promise Joseph Brodsky made when he first introduced Platonov to his American readers, are predicated on the assumption that the socialist context and experience have finally become obsolete and, moreover, that reading Platonov has no effect on making that experience any less obsolete.[2] This is a turn of events Platonov would find rather surprising. He would be surprised not only because he had higher hopes for socialism, but also because he thought his literary work stood in a fundamentally different relation to reality, one that can be defined as broadly realistic. He would be surprised to learn of the modernist interpretive triumph in accomplishing something even his Soviet censors were unable to do: thoroughly separate his work from the reality he was so determined to bring to reason. But most of all he would also be surprised that even at the end of history, with so little seemingly at stake, the methodological habits of those studying his works remain entrenched, his critics still unwilling to acknowledge what ostensibly should no longer matter—the socialist character of Platonov's prose.[3]

A challenge

How can a field of inquiry remain under the spell of a tendency that has long outlived the conditions and the utility of its original formation? This is a question Philip Ross Bullock comes very close to asking in a remarkable recent study on what he sees as the "exclusive emphasis on modernism" in Platonov scholarship. Here I turn briefly to Bullock's 'Platonov and Theories of Modernism' in order to introduce the challenge I hope to address in this chapter. By documenting the exact circumstances under which purely formal considerations of Platonov's work had congealed into a dominant interpretive framework, Bullock's seemingly conventional overview of the main currents in Platonov scholarship begins to draw readers' attention away from the text and towards the ideological and institutional forces shaping its critical reception (301). As Bullock's own anodyne language of intellectual "genealogy" gives way to a sociology of literary criticism, we begin to see that behind all the emphasis on modernist form lie political, social, and institutional exigencies—in short, social content—to which the field responded with great fealty. The exigencies Bullock identifies are almost entirely those of the Cold War: institutionalization of modernism in American academic and publishing markets (301); politicizing of art in establishing claims of greater intellectual freedom (303); overtly political publication patterns (304–305); falsifying of biographical information so that "despite [Platonov's] frequently stated wish to be a *Soviet* writer, his works were repeatedly interpreted as *anti-Soviet*" (302), and so forth. Most of these conditions and patterns have since changed and yet, as Bullock observes almost in disbelief, the exclusive emphasis on modernism remains. How can this be? Since the question is not posed explicitly (and no direct answer is offered), the reader may be tempted to give up on Bullock's argument at this point and conclude that it was only a ruse, an elaborate idealist scheme dressed in the language of literary sociology that sought nothing more than to free Platonov scholarship of its historical and social determinations. This is certainly a possibility but there could also be something more cunning at hand.

Is it not also possible that Bullock is hoping *not only* to draw our attention to the ideological and social forces shaping the field, *but also*, by showing the usual forces to have exhausted themselves, to question whether our grasp of these extra-literary forces is adequate to understanding and breaking through the seemingly paradoxical situation Platonov studies is in today? Is it not possible that the logic governing the field is an expression of more expansive theoretical and political commitments, of aspirations that are yet to be met? That something else had entered the field in those early formative moments, something that remains hidden to this day and that cannot be exorcised by attending only to the forces and agendas that lie beyond the text? Something that is discernible less as a residue of the social within the textual (although it certainly is that) than as a social effect produced directly within the text? Perhaps what we are looking at here is not a concession to textualism but a challenge to literary sociology. A challenge that I believe is twofold: to make visible the more subtle ideological patterns and circuits of influence that have thus far escaped the attention of revisionist scholarship in Slavic literary studies; and to do so by learning to recognize other ways in which political interests and affiliations are conveyed, 'impressed' into the language of criticism rather than openly 'expressed' or, as Brennan may put it, the ways they are explicitly not made explicit.[4]

This is the challenge I hope to take on in this essay, but not before we pay one last debt to Bullock. For in addition to suggesting that a conduit remains within the language of criticism where entire arguments can move freely and imperceptibly, Bullock also provides an unusually strong hint of where we may want to look for an example of hostile theoretical and political content concealing itself in linguistic form. Twice in the course of an otherwise concise history of the modernist tendency, Bullock turns to one particular text—the 1984 version of Joseph Brodsky's 'Catastrophes in the Air'. Here is a work, Bullock tells us, that belongs to a bygone ideological era and yet continues to exert undue influence over the field, where it is often cited but rarely subjected to critical analysis. What gives 'Catastrophes' such power? What authority speaks through this text and keeps it from becoming obsolete?

Victims of language

To turn to Brodsky's text in earnest, we need to set the stage and rehearse what is perhaps already familiar to the reader. The year is 1973. Brodsky is closing the first year of his exile in Ann Arbor. Another decade will have to pass before 'Catastrophes' sees the light of day, first as a lecture at the Guggenheim (1984) and then as an essay in his collection, *Less Than One* (1986). Carl and Ellendea Proffer, who earlier met Brodsky in Leningrad and helped him come to the States (with the promise of a teaching position), are about to publish the first American translation of Platonov's *Kotlovan* (*The Foundation Pit*) at their proudly apolitical Ardis publishing house. The bilingual edition is in need of a preface and Joseph Brodsky, in Russian, pens the first few pages of what eventually becomes the essay Bullock brought to our attention. Those who read Brodsky's prefatory remarks were famously warned that:

> Platonov speaks of a nation which in a sense has become a victim of its own language; or, more precisely, he speaks of this language itself—which turns out to be capable of generating a fictive world and then falling into grammatical dependency on it. It seems to me that therefore Platonov is untranslatable, and in one sense that is a good thing for the language into which he cannot be translated.
>
> (Preface xi)

That a statement like this would make immediate sense to the 'linguistic turn' theorists and right-wing ideologues alike is a testament to a time when political affinities were concealed by aesthetic differences. In this respect, looking closely at what Brodsky amended while rewriting his original passages in English is most instructive—for example, adding a phrase like "revolutionary eschatology" would indicate that by 1984 Brodsky appears to have full command of the conservative American idiom ('Catastrophes' 283, 287). This, however, is not the only—or even the most reactionary—language that speaks through Brodsky's essay. We will have to listen to the text much more attentively in order to hear what, as the old joke goes, 'sounds much better in the original German'. The German I have in mind, as it will soon become clear, is that of Martin Heidegger.

Brodsky's introduction to Platonov begins in a fairly predictable way. The four opening moves can be sketched as follows:

1   "Platonov writes . . . in the language of a greater proximity to New Jerusalem", that is, in the language of "paradise's builders . . . or diggers" (285–286).
2   What is being built, by whom or where—the precise location of this New Jerusalem—does not seem to matter, for "Paradise is a dead-end [*tupik*]" that no thought can scale and no words can describe (Preface ix).
3   The meaning of Platonov cannot thus be ascertained from the specific content of his language. "An aura of arbitrariness" envelops his words (ix); the "ground . . . slips from under" them, leaving the reader "marooned in blinding

proximity to the meaninglessness of the phenomenon this or that word denotes" ('Catastrophes' 287). The particular is of little concern when the analysis is of "this or that," or when, in yet another characteristic move, the reader is told that she can "cast a sort of myopic, estranged glance at any page" and "get a feeling of looking at a cuneiform tablet" (288).

4   Having declared meaning to be "arbitrary," Brodsky turns to Platonov's "writing technique" in search of an interpretive key (287). Through "inflections" (289) and syntactical "inversions" (Preface xi) of the already highly "synthetic" Russian language (x), Platonov "drives language into a semantic dead-end" (ix), producing "concepts totally devoid of any real content" ('Catastrophes' 289). What was denied on the level of content—lest we succumb to some "mimetic" faculty of art (278)—is now readily granted in consideration of form. Thus one reads that "Platonov was able to reveal a self-destructive, eschatological element within the language itself, and that, in turn, was of extremely revealing consequence to the revolutionary eschatology with which history supplied him as his subject matter" (287). One should note that "technique" ceases to be a mere stylistic device and instead becomes the formative principle that governs the whole work. "Technique here is something absolute," as Lukács puts it in 'The Ideology of Modernism' (18). It does not only "reveal" *what* content cannot reveal; it also "reveals" *that* content cannot reveal.

The turn

This is where Brodsky's argument takes an unusual turn. His contempt for any "imitation of reality" ('Catastrophes' 278) is so severe that he cannot simply agree with fellow exiles in characterizing Platonov's deformed language as an "accurate reflection" of the deformed Soviet reality (Geller 402).[5] This would be too profane—a sleight of writer's hand that could fool Soviet censors, but not Brodsky in his pursuit of pure form. Since there can be nothing mimetic about art, form too cannot imitate reality. Against representation, Brodsky offers a view of art that is seen as "creating, or better still, reaching for [reality]" ('Catastrophes' 278). At first this may appear as a careless slip since Brodsky is referring here to two different realities: a reality that is "created" (a new reality, by definition) and another reality that art "reaches" for (a reality that is already present but not 'at-hand'). Another way to say this is that (art reaches for) a reality (that) is not the reality (art creates), or simply that reality is not reality. In a few lines Brodsky will repeat the self-contradictory form of this claim in a manner too explicit to ignore: "Art is not about life," Brodsky explains, "because life is not about life . . . [A]rt, like life, is about what man exists for" (278).[6] The strange turn of phrase is again predicated on the shifting nature of his terms—in this case, the double meaning of "life" that is not "life"—but this time the seemingly untenable proposition is followed by what amounts to an explanation: a peculiar characteristic of man's existence that is always projected beyond itself into the realm of possibilities, into the future, into what is not. The essence of such "life" is that it has no essence ("life is not about life") and that its essence is in its existence ("life is about what man exists for"). Since the category of "what man exists for" is meant to explain both what "life" is and what it is not, it cannot refer to anything that is in any way present in "life." An authentic "life [that] is about what man exists for" would become an inauthentic "life [that] is not about life" at the very moment when "what man exists for" were to actualize itself in history. This means that man cannot "exist for" anything in particular, or, to put it more properly, that "man exists for" *nothing*. "What man exists for," therefore, designates a state of projection—what Heidegger called *Entwurf*—a life that is outside of itself, permanently extended in the world and temporally ecstatic.

One of Brodsky's two realities from above refers precisely to this ontological condition. This is the reality art "reaches," a more fundamental reality that remains present—to poets and within language, as we shall see below—even when obscured by the immediately given or ontic realities 'at-hand'. To put it more accurately, the ontological reality which designates the nothingness of being reveals itself through its absence in the world where lesser (ontic) realities give way to one another. Brodsky can thus speak of art as both "creating" and "reaching for reality" not because he has the same confused reality in mind, but because the process of "creating" a new reality reveals the finitude of ontic realities and, in this nothingness, "reaches" for the more fundamental ontological reality. One should keep this logical pattern in mind, for we will come across it repeatedly.

The proletarian writer in the house of being

Brodsky's argument about Platonov is beholden to Heidegger in three additional ways. In terms of methodology, both the philosopher and the poet rely heavily on discursive manipulation to simulate historical concreteness, logical coherence, and even moral courage. One can say that their method *is* their language. There are, of course, real differences in idiom: Heidegger, a bad poet, relies on the mystification of language and the ringing sound of materialism stemming from conceptually hollow words while Brodsky, a bad philosopher, allows formal properties of language to take over his argument. A poet who "trusts the intuition of language more than his own," Brodsky explains, is "repaid hundredfold" by language when "its subordinate clauses . . . carry him further than his own original intentions or insights would have" (278). Stylistic differences notwithstanding, the basic methodological orientation remains the same, since both Heidegger and Brodsky believe that truth still dwells in language, the oft-cited "house of Being," where it somehow remains sheltered from the oblivion of Being (*Seinsveressenheit*) that permeates our age (Heidegger, 'Letter on Humanism' 217, 242–243). One can account for this extraordinary power of language to withstand history—including the history of its own (mis)use in the hands of realism—by locating within language the same essentially ecstatic structure we saw above in Brodsky's notion of "life" or in Heidegger's Dasein.

In terms of language, this simply 'means' that meaning is always outside and already ahead of itself, that it is 'always already' other than itself, and that this feature is somehow structurally determined. Under the ontological regime where futural possibility is seen as logically anterior, language becomes simultaneously grounded in and undermined by the ambiguity of its own rhetorical modes and tropes. Such outwitting of history by an inner dynamic mechanism points to an agency within language, a playful mind of its own or, as Brodsky put it, "the intuition of language" ('Catastrophes' 278). To hear the silent call of this agent means to submit to the radical undecidability of meaning and, by doing so, to allow for the disclosure of truth within language. Put more provocatively, homelessness (*Heimatlosigkeit*)—dwelling in the aforementioned *Seinsveressenheit*—ends with man's return to language, and not with the building of a foundation for a new home.[7] And this, not to get ahead of our argument, is the obvious and only point Brodsky wants his readers to take from *The Foundation Pit*: the absurdity of any human endeavour to overcome the historical and material contradictions that Brodsky mistakes for second nature. To read *Kotlovan* as a simple allegory of communism—where Platonov's diggers can never dig deep enough to create or even fathom a base stable enough to support the superstructure they know only from slogans—is to willingly misread the text that even in its opening and closing paragraphs directly repudiates Brodsky's take (and his entire hermeneutic) *and* to in fact know nothing of one's own, supposedly erudite references through which most of the jabs are delivered.[8]

The second ontological assumption hidden in Brodsky's argument pertains to the internal temporal structure of "life" understood as Dasein. Heidegger's own writing is soaked with time so that even his nouns routinely appear to be in motion, forgoing whatever conceptual content they may otherwise have. Brodsky's piece on Platonov is less explicitly concerned with time than it is with language—the "first victim of any talk about Utopia" (Preface ix)—although the elegizing tone of his opening paragraph along with its condemnation of eternity could trick his readers into believing that the poet is also mourning the *passing* of time. This is a ruse and yet another example of their language getting ahead of their arguments. As we have seen above, Dasein is a type of being that never has its own being ("life is not about life") or, and this amounts to the same thing, a type of being that always has its own being to be ("life is about what man exists for"). As such, Dasein is to be fundamentally understood in terms of its temporality, as a being "always already" projected into the future. "The meaning of Being [*Sein*] of that being [*Seienden*] we call Dasein will prove to be temporality [*Zietlichkeit*]," Heidegger announces at the beginning of *Being and Time* (17). Such ontologizing of temporality is a seductive proposition until one realizes that not only is the present never present, but the future—abstractly present to an individual Dasein—can never be actualized if Dasein is to exist authentically. The problem here is not that Dasein is made to be finitely transcendent or that it is radically severed from itself (neither of these claims is solely Heideggerian in nature). The problem, instead, is that the future can never come and that this suspension of time is celebrated as authenticity and even freedom. This brings us closer to the third ontological premise in Brodsky's argument—a particular understanding of politics inherent in this highly problematic take on temporality.

It would be tempting to conclude here that time slips into eternity and ceases to be. This is true to the extent that Heidegger inscribes temporality within the existential structure of Dasein and thus renders it (time) atemporal. But that is not all he does. Within this ontological structure of existence, a remnant of a different kind of time survives in the form of Dasein's constant "falling" (*Verfallen*) and continual "thrownness" (*Geworfenheit*) into the world (*Being and Time* 175, 179). Along with "projection," which has been our sole emphasis thus far, "falling" and "thrownness" are the other two basic formal indicators of the ontological structure of Dasein, or what is more commonly known as the *existenzials*. For Heidegger, these are the three primary ways to understand Dasein in the world, from which he can then extrapolate the unitary ontological structure within Dasein. In the context of this triad, "thrownness" designates the state in which Dasein "always already" finds itself in a world that is not of its making and that was, thus, "there" (*da*) before Dasein came into it. We see now that "*da*" in the word "Dasein" is supposed to refer to this "thrown" existential characteristic of a human being within time—indeed, the most basic *existenzial* from which the other two are derived—but if we look beyond the logic imbedded in this common German word for existence, we will notice that the meaning of time has just shifted. The temporality of the world into which Dasein is "thrown" is not the same as the temporality ontologized within Dasein's own internal structure. To put this in more conventional terms, the time of history is not the time of Dasein (the so-called "historicity"), nor does it become so by simply claiming that the nature of Dasein is (to be) historical.[9] The future may yet turn out to be the expression of Dasein's becoming, but the past and the present, the actual world into which Dasein is "thrown," belong at least "initially" to an altogether different understanding of time in which Dasein has already become its own negation, i.e., a thing. It appears that "prior" to the structural "priority" of becoming (historicity) is the "factical" reality of reification (history). If the disjunctions we have encountered thus far were explained in terms of time, this last disjunction opens up *in* time and thus appears to threaten the ontological foundation of the whole project. The answer to the question of disjointed time, the question of why there is always and only history of "thrownness" where historicity of "projection" should stand—a version of that other famous question of why there are beings/things [*Seiendes*] instead of nothing—lies for Heidegger in the "falling" or fallen nature of Dasein. The precise way in which this third *existenzial* closes the gap between two temporal registers will make the politics of ontology explicit in the sections that follow.

First political corollary: fallen nature

The fallen state of Dasein simply means that, in Heidegger's words, Dasein "has initially always already fallen away [*abgefallen*] from itself and fallen prey to the world [*an die Welt verfallen*]" (*Being and Time* 175). In just a few lines,

Heidegger will "define" this kind of being-in-the-world as "inauthentic," and will then proceed to reassure his readers that "not-being-its-self functions as a *positive* possibility" and that it "must be conceived as the kind of being of Dasein closest to it and in which it mostly maintains itself" (175–176). The first thing to notice in this passage is that Dasein's "falling away from itself" is described as an act synonymous with its "falling prey to the world." Inauthentic Dasein "disowns" itself of its ecstatic essence by "absorbing" itself in the world, i.e., by mistaking itself for a thing-like presence whose essence is stable and whose existence is completely determined. However, the same sentence that establishes that there is essentially only one "fall" (to "fall away" is to "fall prey") also introduces an important difference between the two expressions of the "fall." As we have come to expect from Heidegger, a wordplay at the centre of this sentence (*"abfallen"* turns into *"verfallen"*) adds an ominous sense and a whole new factor to the latter ("falling prey to the world") that is simply absent from the former, more matter-of-fact expression of "falling away from itself." In spite of the fact that both expressions refer to the same "falling" phenomenon and the same "falling" culprit, there is already a sense that the "fall" is simultaneously precipitated by two different sets of circumstances: on the one hand, Dasein's own internal constitution and, on the other hand, Dasein's inability to resist what Heidegger later calls the "temptation" of being-in-the-world (177). We are already beginning to see that the "falling" *existenzial* is made to straddle the ontic-ontological difference as a phenomenon that simultaneously pertains to the temporality of history, in which Dasein is literally victimized by its existence in the world, *and* to the temporality of historicity, wherein the "fall" is an inevitable outcome of Dasein's constitution.

One can also see in the above passage that Dasein's primary everyday mode of existence is inauthentic. Dasein is said to have everywhere and "always already" fallen in the world. This "fall," however, is not the type of event where a being falls from one state into another or where it becomes what it is not. Instead, this fallen being is and remains in the fallen state: it "falls" by remaining fallen. To complicate this further, this being "falls" from the essence it does not have in its "everydayness" by remaining in the world into which it was thrown. This means that no "fall" actually takes place. No separate act is needed here to precipitate the "falling" of Dasein because existence itself—all of it—is a priori fallen. Inauthenticity becomes an ontological category in other words. One has to wonder here about Heidegger's duplicitous use of active phrases ("falling" or "remaining in the throw") to describe a foregone conclusion of fallen existence. In a moment we shall return to this but, first, we should point out the first and more obvious political corollary. Heidegger sees ontologizing of inauthenticity as an act of radical historicizing of the human condition, an antidote to the myths of prelapsarian purity and the prophecies of redemption—the "dead-ends" in Brodsky's terminology. If Dasein were only to realize that to exist is to fall, Dasein could then get on with what Heidegger calls—in a remarkable turn of phrase—the "business" of existence (289).[10]

Whatever the "business" of life may be, the "business" of fundamental ontology turns out to be the exact opposite of what Heidegger claims. The emancipatory reduction of life to the finitude of time does not free humanity from myths; it, instead, submits Dasein to the principal myth of *our* time that conceals economically and historically determined forms of alienation behind ontological categories. In fact, in a never-ending game of appropriation of intellectual history that is itself modelled on the "business" of extraction, Heidegger explicitly redefines "alienation" (*Entfremdung*)—a term of distinct Marxian and Hegelian provenance—as an extreme manifestation of Dasein's fallen constitution (178). But this ontologizing of alienation does more than conceal the historical contradictions that produce alienation. We will not exhaust the political implications of Heidegger's temporality without also examining the second dissimulating principle at work here, i.e., the empty gesture of transcendence imbedded in his ontology. If we return to the passage on "falling," we will see that we are well on our way to understanding why inauthenticity can at once be ontologized and transcended in Heidegger's system.

Second political corollary: empty transcendence

We left our discussion of "falling" by noting the odd rhetorical choice of employing active terms to describe something that Dasein should not be able to act on. In asserting that Dasein's inauthenticity consists in its actively remaining in a fallen state, Heidegger seems to once again suggest that the "falling" *existenzial* is more than a function of Dasein's essential, ontological structure. As a historical being-in-the-world, Dasein is depicted as actively inactive not because it can in any way alter the ontological truth of its constitution, but because Heidegger's system requires every ontic Dasein to assume responsibility for its fallen, ontological state. This is Heidegger's way around the problem of mediation, a philosophical approach-cum-ethical principle that can be best described as an affirmation of negation.

The negative thrust of Heidegger's fundamental ontology should be sufficiently clear by now. Reality is a negative category in Heidegger just as it is through a good portion of Western religious and philosophical tradition. One can even say that negation is the key theological insight carried over into all philosophy worthy of the name after the advent of bourgeois ideology. This point is oversimplified but nonetheless critical to recognizing that Heidegger intentionally makes his mark within the tradition of negative philosophy—a point Lucien Goldmann made many years ago. His philosophy stands apart from negative philosophical tradition less in what it claims than in what it completely omits: a concept of transcendence. No transcendence is to take place *in* history and, most certainly, no transcendence *of* history is allowed considering his well-established impatience with traditional metaphysics. At this point one has to be careful not to jump to the obvious conclusion that the place of transcendence of fallen reality is taken by its affirmation, a reversal akin to that carried out by Nietzsche. In Heidegger, affirmation does not so much supplant transcendence; it, in fact, becomes transcendent.

To fully grasp the affirmative moment in Heidegger's argument, one needs to return one last time to Heidegger's passage on "falling" and notice that the fundamental inauthenticity of (all) being is introduced not only as an "initial" condition, implying that some other condition is to supersede it, but also as a necessary one. Heidegger clearly gives the impression that some process is in place whereby humanity can move beyond its fallen existence and attain authenticity. This language is again misleading, for no such process exists in Heidegger, not even in the most ordinary religious sense. Instead of an actual process of overcoming—a process we can describe as a negation of the negation that is inauthenticity—Heidegger offers a flattened view of transcendence paradoxically understood as an affirmation of what exists. This is presented as a view antithetical to positivism because "what exists," ontologically speaking, is nothing. Although Heidegger's use of the term "transcendence" seems to vary at first, its basic affirmative character persists in all of its formulations. Transcendence begins as a constitutive feature of Dasein, something we already encountered when discussing Dasein's "transcendental finitude" in terms of its ecstatic essence. In a certain sense, this type of transcendence has already occurred and Dasein is simply living through the fallout as a being transcendent by nature but forever late to transcend. The other type of transcendence, the type that ostensibly takes place within the ontic or experiential realm, turns on Dasein's readiness to accept alienation as an ontological category. The precise mechanism of this second type of transcendence consists of Dasein coming to "understand" its internal structural relationship to nothingness (*Being and Time* 286) and, by virtue of this understanding, "choose" itself as a being fundamentally divided against itself (287)—a being-guilty/indebted (*Schuldigsein*) in Heidegger's terms (288). Even if only fleetingly, this psychological drama leads to an ethical position of resoluteness in which every individual Dasein is capable of attaining the authentic mode of existence by affirming—not overcoming—the ontic-ontological difference. As was the case with Dasein's essence, authenticity manifests itself as a possibility that is ever present to Dasein and yet has no impact on the absolute triumph of inauthenticity in life. This is so by design, for both inauthenticity and authenticity must exist side by side in what I see as the basic bargain in political ontology: actualization is understood as a betrayal of Being in history while ethical thinking becomes transcendent in its affirmation of the chasm of modern subjectivity. To put it simply, politics is by definition a failure that led to the present condition; ethical posturing is a solution that can lead us out of it.

Lessons in modernism

The purpose of our excursus on the ontological underpinnings of Brodsky's argument was to help make sense of what must have appeared in 1973 as a strange variation on the common theme of celebrating formal experimentation as an expression of artistic freedom and a mark of resistance to the power of the state. Following his analysis of "technique" as a formative principle in Platonov's work, Brodsky suddenly turns away from the obvious Cold War context to declare that "what Platonov was doing with the language went far beyond the framework of that specific utopia" ('Catastrophes' 288). Brodsky even admits that Platonov was "not at all" an "enemy of this utopia"—a remarkable statement notwithstanding its measured, double-negative formulation (288). Brodsky did not grow weak in his modernist conviction or in his rejection of Utopia; on the contrary, he was about to "teach" other critics "a lesson in modernism," a lesson that haunts Platonov scholarship to this day (294).

Brodsky's lesson begins simply enough: one cannot cheat one's way into modernism by hoping that a formal reproduction of deformed Soviet reality can "produce a sufficiently surrealistic or absurdist effect" within a work of art (296). The problem with this half-measure goes beyond the "stylistic conventionality of the [still realist] means of depiction" and it cuts to the core of what the more overtly political and thus less dangerous critics misunderstand about modernism in their eagerness to vilify the Soviet reality (296). Modernism requires complete "aesthetic detachment" from life (271), which means that the "absurdist effect" has to originate from within the artist and not from "his actual historical experience" (273). The same applies to Brodsky's other modernist criteria or the "effects" one is to encounter in a modernist text: "fragmentation, incoherence, an absence of content, a dimmed or bird's-eye view of the human predicament," and so forth (296). This is not a standard solipsistic or idealist position, for Brodsky's artist is very much in a "conversation" (278) with the world around him, cast in the same "unprecedented anthropological tragedy, a genetic backslide whose net result is a drastic reduction of human potential," including his own (271). Brodsky's modernist artist even has a "job" to do "vis-à-vis his society," and this job consists of "projecting" his modernist "sensibility . . . as the only available route of departure from the known, captive self" (273). "If art teaches men anything," Brodsky continues, "it is to become like art [presumably absurd, fragmented, incoherent, and without content], and not like other men" (273). Although Brodsky's artist is thrown into the world, the truth he is communicating is not of that world. This is possible, he asserts in an oracular fashion, because "art has outlived tragedy" (273). We have to briefly pause here in order to explain Brodsky's strange turn of phrase that, in fact, is yet another encoded provocation.

In Brodsky's use, "art" refers to modernist art in Russia and then only to two genuine modernist writers, Dostoevsky and Platonov. All other art and all other writers "went down the . . . path of mimetic writing, and at several removes [from Tolstoy] have reached the pits of socialist-realism" (277). Brodsky's basic distinction here is between (modernist) art and (realist) "social anthropology" (269), and if there is any doubt about what he thinks of the realist "world view" that "guides . . . a writer and a policeman" alike (275), the following paragraph should settle it:

> Russian prose quickly deteriorated into the debilitated being's flattering self-portrayal. A caveman began to depict his cave; the only indication that this was still art was that, on the wall, it looked more spacious and better lit than in reality. Also, it housed more animals, as well as, tractors.
>
> (270)

It is clear what Brodsky considers "art" in the above provocation, but what does he mean by "tragedy" so that "art" can be said to "have outlived" it? "Tragedy,"

Brodsky explains, "is history's chosen genre," and if it were not for "literature's own resilience," "tragedy" would have been history's only genre (271). "Tragedy" seems to refer to a certain inevitable quality of history that is "always already" fallen, fated, or simply "tragic" in the most banal, non-literary sense of the word. When left to its own devices, i.e., whenever left simply to human beings, history finds its inner meaning and its truth expressed in a tragic form. Considering that the obvious historical "tragedy" he has in mind is the "national upheaval that took place in Russia," and that he has already referred to this "catastrophe" in some dozen or so equally suggestive ways, one must wonder why Brodsky suddenly finds it necessary to introduce yet another term—a term with a distinct literary and philosophical meaning—particularly when he shows no interest in using the term accurately. In other words, what does the word "tragedy" convey about the Soviet "totalitarian present" (270) that the sentence about the "vast, corpse-strewn, treachery-ridden historical vista, whose very air turned solid with howls of ubiquitous grief," does not (280)? A rather bizarre answer to this question presents itself in an offhand comment Brodsky makes about "Aristotelian art-to-life proximity" on the next page (272). "Tragedy" is Brodsky's way of alluding to Aristotle's *Poetics* and the privileged status granted not only to tragic literary form, but also to the mimetic principle in art from the very beginning of literary criticism in the West. "Tragedy" thus refers both to the catastrophic nature of history and to the mimetic quality in art, which, in Brodsky's mind, are two intertwined manifestations of the same problem he dubs "religious humanism" (254). To say that "tragedy is history's chosen genre" simply means that realism is an art form of fallen history. Fortunately a higher, modernist form of art has "outlived" the never-ending tragedy of history/realism. This is less a statement about modernism's unlikely feat of survival in the hands of "history and her ironclad agent: *Polizeistaat*" (279) than an indication of a different kind of temporality that applies to modernism.

For art to outlive tragedy, the artist has to "retreat back into" his medium and retrieve from it what tragedy could not spoil (287). But what can that possibly be in the case of the Soviet writer who so willingly submitted himself to "newspeak" (288)? What deeper truth can survive within language if this language was already shown to be the first victim of Utopia (Preface ix)? Clearly no truth can be retrieved from its content, and none from its form as long as form still holds on to its own historical and ideological content. And yet, if we are to ask that last question anew, in good Heideggerian fashion, we will realize that the answer was already contained within the question. The self-revealing truth of the fallen language of Platonov is that of the inherent failure of language. This truth can "outlive" the tragedy of history because it is the truth of history unknown to itself. Thus Platonov is said to have "driven [his] language into a semantic dead-end" (ix) so that in this "relentless and implacable absurdity" ('Catastrophes' 287) he could "reveal *in language itself* the philosophy of dead-end" (Preface ix; emphasis added). In other words, a formal device that creates the "absurdist effect" on the level of a literary work discloses the "always already" absurd formal quality of language itself.

More troubling from our perspective is what Brodsky does with this "modernist" truth or to what purpose he turns the Soviet writer against language—"the real target" of Platonov's works, as Brodsky tells us ('Catastrophes' 283). Why would the exiled literary critic, himself a poet and a Nobel Laureate, insist on the "fallenness" of language? We get a hint of a possible answer when Brodsky declares that "the presence of the absurd in grammar says something not just about a particular tragedy, but about a human race as a whole" (Preface ix). What could the absurdity of language possibly "say" about the human race? According to Brodsky's increasingly deviant scheme, the modernist writer is able to reveal the fallen nature of language *in* language or, more precisely, he allows for the truth of language's fallen nature to reveal itself *in* language and, by doing so, demonstrate in the failure of language a higher truth *about* language. This is to say, he points to a purer form of language's existence, a form that has absolutely no content and, as such, cannot be subject to the ruination of history. It is in this new form that art "outlives" history, and this historical miracle, this ahistorical historicity of literature, is, I submit, what Brodsky is really after. Platonov matters to Brodsky only to the extent that he can use his modernist analytic to prise open a disjunction in time and arrive at this new temporality. In Heidegger's terms, the modernist writer is conceived here as the new ethical self who hears the silent ontological truth that still dwells in ontic language and understands this truth—the oldest truth that has to appear anew to the epoch that took the possibility of transcendence a bit too seriously—to be that of a purer, higher form of existence devoid of essence and any possibility of transcendence.

I will finish with an analogy: to say that the absurd in literature reflects Soviet reality is analogous to taking a shot at communism from a conventional weapon and badly missing the target. Brodsky realized this and thus went for what can only be described as a 'nuclear option' considering its consequence on "life": the absurd effect in Platonov's prose corresponds to *nothing* in particular—it has no content as both he and Heidegger are fond of saying—and as such, it expresses the *nothingness* of all existence. This is a strike from which communism as a dialectics of hope cannot recover.[11]

Notes

1 The phrase is Thomas Seifrid's (*Uncertainties of Spirit* 176). In an early attempt to bridge the divide between the "two Platonovs," Seifrid argued that Platonov was *neither* an effete modernist (his assault on linguistic conventions was not self-conscious) *nor* a socialist realist (here Seifrid relies on Boris Groys's "style and a half" thesis; see Seifrid's 'Platonov, Socialist Realism, and the Legacy of the Avant-Garde' 242). The "two Platonovs" are reconciled under a new category of "*de facto* modernism" where the remnants of realism are read as "ontologically vivid parodies of the [utopian] genre in which the fusion [between the transcendent and immediate reality] routinely fails to take place" (*Uncertainties* 18).

2 I am referring to the 2014 special section of *Slavic Review* and, in the above quote, to Eric Naiman's comments about the state of Platonov studies and its "anxiety about relevance" (804). "Obsolete" in this case are the "paradigms" Platonov fought against. The immediate Soviet context can still provide some general bearings, but not the

irrevocable horizon of meaning, lost to the ever-modernizing academic markets. "New paradigms" and "interlocutors" are called forth to demonstrate that Platonov can speak beyond his age. According to the editor of the special edition, Nariman Skakov, Platonov's work of course "belongs to the symbolic order of the era" and yet the truth it speaks—mediated by such non-antiquated notions as Aristotle's distinction between *zoē* and *bios* or the status of *homo sacer* in Roman law—takes a rather banal form in the contemporary jargon on undecidability: "Platonov . . . found himself in the process of constant oscillation between various binaries . . . but he managed to evade synthetic mediation" (726).

3 Largely ignored in the organs of official scholarship, a handful of critics have insisted on seeing Platonov's work in a different light—not against the grain or beyond the horizon of the Soviet experience, but as a work of "socialist literature," giving expression to an emerging socialist consciousness and a communist culture (Jameson 73). Apart from Fredric Jameson's ruminations on the changing utopian form in *Chevengur* (*The Seeds of Time* 90) and Slavoj Žižek's notes on Platonov's rejection of millenarianism and post-humanism (*Living in the End Times* 374–375), see McKenzie Wark's take on Platonov's "proletarian writing" as a series of *new* historical novels—in line with Lukács's definition at the end of his 1937 *The Historical Novel* (*Molecular Red* 116, 370); and a *tour de force* article by Artemy Magun, who sees Platonov preserving in his dialectic of solitude and community the "constituent (or *destituent*) power of communist revolution" ('Negativity in Communism' 19).

4 This entails a shift in literary sociology that Brennan defines in the following way: "What seems overlooked in much of our criticism . . . are the benefits of objectifying theoretical style, seeing its ideological gestures as formal components in and of themselves. The first thing to note is that this involves more than a purely formal analysis. It demands a description of the political logic of form as well as an account of how formal devices are often homologies of social agendas. Fredric Jameson's idea of 'cognitive mapping' is an example of such a homology between *socius* and form, grafting the former onto the latter. But is there a way of getting to the reverse by returning, paradoxically, to the text in order to examine from the outside the rhetorical patterns that betray a political significance never made explicit . . . in fact, explicitly not made so?" ('Running and Dodging' 278).

5 In his 1972 introduction to the first (Western) edition of *Chevengur* and then again in his 1982 book, *Andrey Platonov v poiskakh schast'ia*, Mikhail Geller follows the same steps I outlined above and even uses some of the same language we saw in Brodsky: e.g., "the language of utopia cripples the human being" [*iazik utopii kalechit cheloveka*] and "forms out of him an obedient citizen of an ideal society" [*formiruet iz cheloveka poslushnogo zhitelia ideal'nogo obshestva*] (Geller 291; my translation). Beneath these similarities, however, lies a rift in the conservative worldview. As Seifrid has noted repeatedly in response to Geller's monograph, an approach that "attributes the morbidity characteristic of Platonov, not to his art, but to the world it represents" still makes the same assumptions of verisimilitude that one sees in realism (Review 570; *Uncertainties* 16).

6 Brodsky appears to have been fond of this expression. He used it in a slightly different form in his 1979 interview with John Glad, an abridged version of which was published in Glad's 1993 *Conversations in Exile: Russian Writers Abroad*, pp. 103–104.

7 Heidegger's discussion of *Heimatlosigkeit* in his 'Letter on Humanism' is explicitly tied to his critique of the "metaphysical essence of labor" in Marx and Hegel that supposedly sells man short (243).

8 For example, one cannot just note the infernal setting in *Kotlovan* and the protagonist's descent into the abyss of communist folly (Platonov's title and the opening sentence are clear references to Dante's *Inferno*) without also acknowledging that Platonov's divine comedy—not human tragedy as Brodsky will soon claim—is intentionally structured, in line with Dante's work, as an epic of human triumph. This triumph lies not in some

dim image of future redemption which is as invisible to Platonov as it was to Dante, but in the work of immanent critique. That no shortcuts to paradise are possible in a world riven by contradictions is the point of the next sentence in *Kotlovan* where the protagonist, having lost his job in the previous sentence due to his absorption in thought, steps "outside" in order to "better understand his future out in the air" (Platonov 1). The "outside," however, cannot respond. "The situation in nature was quiet" because, as Hegel may put it, the subject can only understand reality by witnessing his own work, i.e., by coming to see himself objectively in his own thought (cf., Brennan, *Borrowed Light* 109, 118).

9   In Pierre Bourdieu's perceptive observation from *The Political Ontology of Martin Heidegger*, "historicity" amounts to the "eternalization of history" in order to avoid the "historicization of the eternal" (63) and as such, it is a prime example of what Bourdieu calls, in reference to Kant's Copernican revolution, "the conservative revolution" in philosophy: a strategy of "radical overcoming which allows everything to be preserved behind the appearance of everything changing" (62–63).

10   Heidegger's statement, "'Life' is a 'business,' whether or not it covers its costs" (*Being and Time* 289), is a reference to Schopenhauer's pronouncement that "nothing whatever is worth our exertions, our efforts, and our struggles, that all good things are empty and fleeting, that the world on all sides is bankrupt, and that life is a business that does not cover its costs" (*The World as Will and Representation* 574). Heidegger had no patience with this despondent toying with debt-forgiveness. Instead he distilled from Schopenhauer the only truth that matters, viz., that "life is a business" and that debt/guilt (*Schuld*) is its essence.

11   The author wishes to thank Timothy Brennan, Robin Brown, Keya Ganguly, Dinara Georgeoliani, Asher Ghaffar, and Alice Lovejoy for reading earlier versions of this essay.

Works cited

Bloch, Ernst. *The Principle of Hope.* Translated by Neville Plaice et al., vol. 3, MIT Press, 1986.

Bourdieu, Pierre. *The Political Ontology of Martin Heidegger.* Translated by Peter Collier, Stanford UP, 1991.

Brennan, Timothy. *Borrowed Light: Vico, Hegel, and the Colonies.* Stanford UP, 2014.

—. "Running and Dodging: The Rhetoric of Doubleness in Contemporary Theory." *New Literary History*, vol. 41, no. 2, Spring 2010, pp. 277–299.

Brodsky, Joseph. "Catastrophes in the Air." *Less Than One: Selected Essays*, by Joseph Brodsky, Farrar, Straus and Giroux, 1986, pp. 268–303.

—. "Joseph Brodsky." *Conversations in Exile: Russian Writers Abroad*, edited by John Glad, Duke UP, 1993, pp. 101–113.

—. Preface / Predislovie. *The Foundation Pit / Kotlovan*, by Andrey Platonov, translated by Thomas P. Whitney, bilingual ed., Ardis, 1973, pp. ix–xii, 163–165.

Bullock, Philip Ross. "Platonov and Theories of Modernism." *Andrej Platonov*, special issue of *Russian Literature*, edited by Ben W. Dhooge and Thomas Langerak, vol. 73, no. 1–2, Jan.-Feb. 2013, pp. 301–322.

Geller, Mikhail. *Andrey Platonov v poiskakh schast'ia* [Andrey Platonov in search of happiness]. YMCA Press, 1982.

Heidegger, Martin. *Being and Time.* Translated by Joan Stambaugh, revised ed., SUNY Press, 2010.

—. "Letter on Humanism." *Basic Writings.* Edited by David Farrell Krell, revised and expanded ed., Harper, 1993, pp. 213–265.

Jameson, Fredric. *The Seeds of Time.* Columbia UP, 1994.

Lukács, Georg. *The Historical Novel.* Translated by Hannah and Stanley Mitchell, U of Nebraska P, 1983.

—. "The Ideology of Modernism." *The Meaning of Contemporary Realism.* Translated by John and Necke Mander, Merlin Press, 1963, pp. 17–46.

Magun, Artemy. "Negativity in Communism: Ontology and Politics." *Russian Sociological Review*, vol. 13, no. 1, 2014, pp. 9–25.

Naiman, Eric. Afterword. *Platonov's Turkmenia*, special issue of *Slavic Review*, edited by Nariman Skakov, vol. 73, no. 4, Winter 2014, pp. 801–804.

Platonov, Andrey. *The Foundation Pit / Kotlovan.* Translated by Thomas P. Whitney, preface by Joseph Brodsky, bilingual ed., Ardis, 1973.

Schopenhauer, Arthur. *The World as Will and Representation.* Translated by E.F.J. Payne, vol. 2, Dover, 1966.

Seifrid, Thomas. *Andrei Platonov: Uncertainties of Spirit.* Cambridge UP, 1992.

—. "Platonov, Socialist Realism, and the Legacy of the Avant-Garde." *Laboratory of Dreams: The Russian Avant-Garde and Cultural Experiment*, edited by John E. Bowlt and Olga Matich, Stanford UP, 1996, pp. 235–244.

—. Review. *Andrey Platonov v poiskakh schast'ia*, by Mikhail Geller, *The Slavic and East European Journal*, vol. 28, no. 4, Winter 1984, pp. 569–571.

Skakov, Nariman. "Introduction: Andrei Platonov, an Engineer of the Human Soul." *Platonov's Turkmenia*, special issue of *Slavic Review*, edited by Skakov, vol. 73, no. 4, Winter 2014, pp. 719–726.

Wark, McKenzie. *Molecular Red: Theory for the Anthropocene.* EPUB ed., Verso, 2015.

Žižek, Slavoj. *Living in the End Times.* Verso, 2010.

7 Aesthetic re-imaginings of Mexican sovereignty

Estridentismo's anti-imperialist avant-garde

Christian Gerzso

[Marxist] criticism requires a new set of categories drawn, as they always have been, from anticolonial terrain: montage, unevenness, vulgarity, sacrifice, and polemic.

(Timothy Brennan, *Borrowed Light*)

The interwar period offered for the first time a true global challenge to the imperial system, as Timothy Brennan points out in *Borrowed Light: Vico, Hegel, and the Colonies* (2014); and, not surprisingly, it is in this context that we see the consolidation of peripheral aesthetics committed to Marxist and anticolonial politics. According to Brennan, these anti-imperialist projects sought to explicitly separate themselves from their European counterparts in their "persistent critique, and even mockery, of literary modernism and the avant-gardes, both as a style and a social outlook" (235). As Brennan explains, the peripheral anticolonial literature of this period was one "opposed to *irony* itself—the inheritance of Vico's and Hegel's little-known philosophical assault on irony" (235). Thus, to the commonly celebrated strategies of dissimulation of high modernism and the European avant-gardes, Brennan opposes the anti-ironic aesthetic outlooks of the global periphery. This is why, in contrast to a focus on irresolution or slippages of meaning, a study of peripheral aesthetics requires a different set of categories drawn from these anticolonial terrains, as Brennan points out in the epigraph above: "montage, unevenness, vulgarity, sacrifice, and polemic" (236). Brennan's dual emphasis on anti-imperialist politics and vernacular forms can open up new paths of inquiry into the cultural production that emerged at this crucial moment of anti-capitalist contestation, particularly in post-revolutionary Mexico since, as Brennan rightly observes, the armed conflict known as the Mexican Revolution of 1910–1920 represented one of these serious challenges and promises of an alternative system (235). In particular, the style and outlook of the Mexican avant-garde movement *Estridentismo*, or 'Stridentism', which has been the subject of increasing attention over recent decades,[1] offers a unique case to test Brennan's conception of peripheral aesthetics.

Estridentismo, together with other instances of post-revolutionary cultural production, such as Muralism or the novel of the Revolution,[2] was indeed distinct from

European avant-gardes in the way in which it imagined alternatives to European economic and aesthetic dominance and incorporated autochthonous, popular forms. Nevertheless, at crucial moments of *Estridentismo*'s production, irony did play a central role in its adversarial and polemical stance. So, on the one hand, this movement confirms Brennan's conception of peripheral aesthetics, which must be taken seriously on its own terms as a challenge to rather than a mere derivation of European avant-gardes. On the other hand, the incorporation of irony, especially in the instances in which it served an anti-imperialist critique, forces us to expand Brennan's characterization and consider the potential of a politics of irony, irreverently deployed against capitalist power from a position of dependency.[3]

The 1920s in Mexico were a moment of intense and contentious institutional rebuilding along the emergent nationalist ideology of the post-revolutionary state. As Ignacio Sánchez Prado (2009) points out, this ideology was crystallized above all in Manuel Gamio's seminal anthropological work, *Forging a Fatherland* (1916).[4] Sánchez Prado quotes Luis Villoro's account of *Forging a Fatherland*, which "expressed the ideas of the Revolution better than any other work from the period: its social nationalism, the search for an autochthonous culture, the improvement of the masses through the conscious actions of a popular state, the redemption of indigenous peasants, and the construction of a more egalitarian society" (22; my translation). Muralism, especially the well-known frescos of Diego Rivera and David Alfaro Siqueiros in the Secretariat of Public Education, the National Preparatory School, and the Palace of Fine Arts, among others, is usually taken as the privileged instance of this post-revolutionary ideology, with its depictions of Mexican history as an epic struggle towards an egalitarian and racially mixed society, relying on figuration and celebrating the peasantry and the working classes. Nevertheless, these murals were certainly not the only instance of cultural production committed to this project. The case of *Estridentismo* reveals how neither this commitment nor the aesthetic forms it adopted were monolithic.[5]

The poet Manuel Maples Arce founded *Estridentismo* when he pasted copies of his manifesto, *Actual No. 1*, on the streets of Mexico City in December 1921. This "vanguardist leaflet," as Maples Arce subtitled his manifesto, fiercely attacked Mexican cultural institutions, including art academies, in the immediate aftermath of the Revolution. In doing so, *Actual No. 1* explicitly borrowed the well-known tropes and aggressive style of Italian Futurism: "Chopin to the electric chair! . . . The Futurists . . . have already called for the murder of Clair de lune in block letters" (Schneider 269).[6] But to this, Maples Arce added attacks on the "rancid" provincialism of Mexican artists and intellectuals, especially those dependent on state support:

> I call upon all the young poets, painters, and sculptors of Mexico, those who have not been corrupted by the government's gold and sinecures . . . who have not been undone by the pitiful and goal-driven compliments of our nationalist milieu, reeking with the stench of *pulquerías* and fried food.
>
> (Schneider 273–274)

To combat this provincial "stench," Maples Arce prescribed an engagement with the international avant-gardes and a staunchly independent stance with respect to Mexican government institutions. In the second manifesto, distributed in the city of Puebla on January 1, 1923, the poets Germán List Arzubide and Salvador Gallardo, who had just joined the estridentista cause, followed Maples Arce's lead. Thus, in its initial phase, which would last roughly until 1924, *Estridentismo* used a destructive form of irony to shock the local elites and cultural establishment, and adopted an emphatically anti-institutional stance that closely resembled that of the European avant-gardes. At the same time, *Estridentismo* incorporated local references and vernacular humour and speech. The second manifesto famously ends with the phrase "¡Viva el mole de guajolote!," or "long live turkey mole!," celebrating one of the main dishes of Puebla cuisine and using the indigenous, Náhuatl name, rather than the Spanish one for 'turkey' (Schneider 277).

Estridentismo's independent stance changed dramatically when Maples Arce joined the ranks of the post-revolutionary regime in 1925. Once he obtained his law degree, Maples Arce moved from Mexico City to the city of Xalapa, in his native state of Veracruz, and soon became Secretary of State to Governor Heriberto Jara. Given Maples Arce's new political duties, he called on his estridentista collaborators to join him in the provincial city and put them in charge not only of the movement itself but also of some of the state's main cultural institutions: List Arzubide directed the state's publishing house, the *Talleres Gráficos del Estado de Veracruz*, as well as the movement's new magazine, *Horizonte*. Governor Jara was a former revolutionary leader who had been one of the framers of the 1917 Constitution and whose commitment to labour rights, particularly to workers in the oil industry, put him on a collision course with President Plutarco Elías Calles during his governorship. This meant that in this period, between 1925 and 1927, which we can label its institutional phase, *Estridentismo* joined one of the most radically egalitarian versions of the post-revolutionary regime, overtly embracing its socialist politics. This also entailed gradually leaving avant-garde irony behind: in *Horizonte*, published between 1926 and 1927, the estridentistas now included propaganda pieces emphatically making the case for nation building. In other words, as Lynda Klich has observed, *Estridentismo* moved from a destructive to a "constructive rhetoric" ('Estridentópolis' 112), thus shifting from fierce irony, and indeed strategies of dissimulation and obscure references, to explicit and emphatic commitment. But whereas scholarship on *Estrdentismo* has pointed out these obvious transformations in the movement, it has tended to treat them as an awkward compromise with the state.[7] By contrast, I argue that these changes represent the emergence of an aesthetic outlook that was constitutively anti-imperialist, even if it remained uneven and incomplete. *Estridentismo*'s collaboration with Jara decisively redefined their project for the span of two years, until the Veracruz governor was removed from office in October 1927, due precisely to his support of oil workers against American and British companies.[8]

Yet, even within this period of institutional support, we find in the work of some of its figures, namely the writer Xavier Icaza[9] and the artist Ramón Alva de la Canal, a different kind of irony and humour from the one Maples Arce, List Arzubide, and Gallardo employed in their early manifestos: one that primarily took aim at Euro-American imperialism and more decidedly incorporated popular speech and music. Icaza would later label this aesthetic "tropical," distinguishing it from the European avant-gardes both in political and formal terms (*Revolución Mexicana* 43). For this reason, in the present chapter I first concentrate on Icaza's articulation of this anti-imperialist aesthetic. Then, I analyse how he develops it in his short novel, *Panchito Chapopote*, published in Mexico City in 1928, but written in Xalapa in 1926, at the height of the estridentista collaboration with Jara. Icaza's narrative, accompanied by woodcuts by Alva de la Canal, is the first fiction piece in Mexico to deal with the oil question; and crucially, as its hybrid form makes clear (the novel juxtaposes an intrusive narrator with popular songs, truncated narrative summaries, and dramatic scripts), its anti-imperialist and vernacular aesthetics are not opposed to experimentation.[10] In other words, not all cases of left, anticolonial aesthetics, even within *Estridentismo*, reverted to figuration or social realism once they began to collaborate with the state. Finally, I approach *Horizonte* as another kind of generic hybrid: a propaganda outlet for the Veracruz government that the estridentistas still viewed as an avant-garde publication. *Horizonte* combined estridentista experimentation in poetry and prose with reports on public works carried out by the state government; articles on the oil question and rural and education reform; pedagogical pieces on Mexican historical figures; translations of modern literary classics by Poe and Chekhov; photography by Tina Modotti and Edward Weston; and reproductions of painting and graphic work by Diego Rivera, Rufino Tamayo, Leopoldo Méndez, and Alva de la Canal, among others (the latter two worked as designers for the magazine). In this way, *Horizonte* represents a crucial instance of *Estridentismo*'s formally uneven yet politically committed aesthetics: one that combined polemics on behalf of a socialist project with some more recognizably avant-garde writing, as Brennan explains in the epigraph at the beginning of this section. At this critical moment of national reconstruction, *Estridentismo* regarded both aesthetic and political writing as complementary means of imagining Mexican sovereignty against foreign pressures.

Anti-imperialist aesthetics

The novel was by all means too lengthy. It needed to incorporate popular forms to a greater degree: not merely cite or borrow *sones* or *corridos*, but use them as an integral part by taking their rhythms, turns, and words. Do not cite the *rumba*: write, in tropical books, with the undulating rhythm of the *rumba*.

(Xavier Icaza, 'The Mexican Revolution and Literature')

The study of *Estridentismo* as a key instance of interwar, anti-imperialist aesthetics follows Brennan's call to recognize the single modernity of capitalism and the place that peripheral cultural production has within it. In other words, Mexican post-revolutionary culture does not represent an incommensurable mode of thinking, untouched by European influence. In the case of *Estridentismo*, this is made obvious, among other aspects, by its explicit borrowing of futurist tropes, such as the celebration of speed and industrial technology. But even as they moved away from these elements in later phases of the movement, estridentista writers and artists explicitly conceived of their production within a global context of capitalist imperialism. This was because they understood well, in Brennan's words, that "modernity is singular because of the overdeveloped and interlocking global systems of capital, always the prime motives of colonialism and imperialism," and not because of a preference for a "single-minded" theory of universality, as we tend to regard characterizations of a single modernity today (13). Still, as Brennan reminds us, the existence of a single modernity poses a challenge: "if in the present stage of colonialism we are all left with this single governing logic, what place is left for the non-Eurocentric?" (13). By focusing on *Estridentismo*, I offer a partial response to this challenge that takes us beyond the confines of European cultural production and demonstrates the awareness by Mexican artists and intellectuals of their place within these uneven relationships, as well as their ability to conceive of aesthetic and political alternatives.

On the one hand, *Estridentismo* demonstrates how Mexican post-revolutionary aesthetics are steeped in this single logic, targeting its global economic patterns, developing its media, such as the avant-garde periodical, and contributing to a reassessment of its aesthetic forms. On the other hand, *Estridentismo* is not merely derivative, or an uneasy "negotiation" between metropolitan impositions and native forms, as scholars tend to depict the movement (Flores 261). In estridentista production between 1925 and 1927 we can appreciate some of the characteristics that Brennan observes in peripheral aesthetics. Many of *Horizonte*'s articles and editorials are fiercely polemical in their attacks against Mexican conservatism and American interference in economic and political affairs. Likewise, the magazine includes pedagogical pieces on popular music, while *Panchito Chapopote* incorporates these forms into its narrative structure in a quasi-cinematic montage of different kinds of discourse (Brushwood, 'Bases' 166–170). This is why we must frame *Estridentismo* within the movement's own conceptions of a specifically Mexican avant-garde, rather than imposing European models of analysis.

Estridentistas themselves sought to theorize these post-revolutionary avant-gardes as unique kinds of anti-imperialist aesthetics. In a lecture delivered at the Palace of Fine Arts in 1934, entitled *La Revolución Mexicana y la literatura* (*The Mexican Revolution and Literature*),[11] Icaza looked back on the vanguardist production of the 1920s.[12] As he contended, the most significant trait in the painting, theatre, and literature he celebrated was the aim to join the political and aesthetic revolution (Icaza, *Revolución Mexicana* 38). In this specific kind of political avant-garde, he included his own writing and that of other estridentistas, to be sure, as well as Muralism and the novel of the Revolution, especially Mariano Azuela's *Los de Abajo* (*The Underdogs*, 1915), which the estridentistas reissued with the *Biblioteca Popular* collection of the *Talleres Gráficos* in 1927. Just as List Arzubide had done in his editorials for *Horizonte*, in his lecture, Icaza conceives the Mexican Revolution as a two-part process in which the armed conflict of the teens would need to be followed by a peaceful phase of institutional reconstruction (Icaza, *Revolución Mexicana* 12, 14). This second phase would fulfil the egalitarian promises of the Revolution in the realm of culture, through, for instance, new methods of rural and technical education. In other words, rather than viewing the relationship between politics and culture as a compromise between two ultimately irreconcilable spheres, estridentistas conceptualized it as a back and forth between actual struggle, institutional rebuilding, legislation, and art making, all aiming towards the same political goal.

This meant that literature for Icaza fulfilled a major role in forging an anti-imperialist outlook, even if, in his view, estridentista literature of the 1920s, as well as the novel of the Revolution, still depicted a chaotic and violent Mexico (*Revolución Mexicana* 12, 33–37). In spite of this lack of a clear political programme, such as the one he saw in Rivera's murals, Icaza insists on the political commitment of estridentista literature, targeting "imperialism and junkerism" (42). Moreover, Icaza does not advocate avant-garde opacity or recalcitrant irony. While he celebrates formal experimentation, he clarifies that his aim in his writing of the 1920s was to craft a popular and vernacular aesthetic (*Revolución Mexicana* 38, 43–44). If, in *Panchito Chapopote*, he sought to "pitilessly shorten the unpardonable dimensions" of the realist novel, his purpose was to create brief, synthetic, and direct images that share the scale, language, tone, and "rough taste" of Mexican popular forms (*Revolución Mexicana* 44). To be sure, this had formal consequences that, in fact, make *Panchito Chapopote* a pioneer of Mexican metafiction: first, as he theorizes in his lecture, this meant that, rather than getting lost in the romantic "entanglements" of its protagonist, his novel set out to present "the masses, society, or the people" as its "hero" (43). Icaza's purpose was to make sense of the "national panorama" in relation to the rest of the world in a "brief and contrasting manner" (43–44). In other words, Icaza highlights his interest in political allegories, rather than in revelling in the everyday details of middle-class characters, as he had done in his first novel, *Dilema* (1921), and his first collection of short stories, *Gente mexicana* (1924). Second, Icaza explains his emphasis on "direct images and short phrases" as stemming from "folkloric sources," particularly popular music and idioms, thus creating a democratic, vernacular aesthetic (*Revolución Mexicana* 38). This is why Icaza does not merely "cite or borrow" popular culture, but seeks to fully integrate it into his writing: "Do not cite the *rumba*: write, in tropical books, with the undulating rhythm of the *rumba*" (43). In this way, Icaza contrasts "tropical" aesthetics with the European avant-gardes that merely cite exotic references. By the same token, this kind of literature does not treat imperialism as a concern just at the level of content: "tropical books," as Icaza defines them, are at once aware of their peripheral condition and able to put forth new, popular forms.

Thus, if we follow Brennan's argument on the single modernity of global capitalism, far from exclusively granting agency to its centres of production, the result of acknowledging the existence of this singularity would involve, by contrast, the recognition of the "original contributions, corrections, and adaptations of metropolitan ideas," or in other words, a "sense of mutuality in global articulations of modern life" (13). Anthony Stanton (2014) has made a similar point with respect to Latin-American avant-gardes: "it has become quite clear that Latin-American avant-gardes represent original movements, clearly differentiated from European ones. Their products, even those that openly assimilate foreign stimuli, have their own distinct marks" (*Modernidad, vanguardia* 28). One of these unique traits, according to Stanton, was precisely their project of bringing together "political and aesthetic revolution," and, significantly, *Estridentismo* "was the first movement in Latin-America" to attempt this (*Modernidad, vanguardia* 31). As I argue, another one of these distinct marks was their increasingly ambivalent recourse to irony. Especially in *Panchito Chapopote*, irony is the mechanism through which estridentistas ridiculed Euro-American imperialism and resisted subordination, while at the same time they acknowledged their position in an asymmetrical relationship. The issue with irony, for the estridentistas, was that its irreverence could lead to disillusionment, and so, at the time they committed themselves to Veracruz's constructive programme, it became less useful. In order to examine the potential and limitations of Icaza's anti-imperialist irony, as well as to test his own later theory of "tropical" aesthetics, I now turn to *Panchito Chapopote*.

History as farce: *Panchito Chapopote*

As a result of John Brushwood's studies of the 1980s, and Evodio Escalante's and Elissa Rashkin's of the early 2000s, *Panchito Chapopote* is now much better known, even if it is still not widely read. But because these studies are few and far between, they understandably insist on the aesthetic value of Icaza's novel, while marvelling at its formal experimentation. This means that they devote significant space to listing its innovative qualities, while leaving less room for the interpretation of these, especially in relation to the frame Icaza provided in his lecture.[13] While these critics rightly point out Icaza's reliance on popular speech and music, this still needs to be further considered in relation to the novel's anti-imperialist politics, which all of these critics appropriately highlight. In addition to these studies, Tatiana Flores (2013) offers a lucid reading of *Panchito Chapopote*, primarily focusing on the relationship between Icaza's narration and Alva de la Canal's woodcuts (Flores 256–264), which accompany both editions of the novel (1928 and 1961). This relationship is crucial to emphasize, since Icaza explicitly conceives his novel as a generic hybrid, and since estridentista writers collaborated intensely with visual artists. Icaza subtitles his novel: "Tropical *retablo* or account of an extraordinary event in the heroic Veracruz." The term '*retablo*' refers to easel paintings depicting narrative subjects, either in altarpieces in churches or in folk art. *Panchito Chapopote* is certainly a narrative of the secular, popular kind.

Icaza structures his novel as a series of more or less short vignettes, interspersed with dramatic scripts. It opens with Panchito, an Afro-Mexican farmer, witnessing a rumba dance in the main square in the Port of Veracruz. Soon afterwards, he follows a prostitute to a brothel, and begins to tell her the story of his enrichment, having sold his "cursed" lands, filled with oil, to an American company. He first explains how he got his nickname ('*chapopote*' means charcoal, hence the moniker refers both to his skin colour and the oil in his lands), but soon the narration pans out to the American and British intervention. Later in the novel, the narrator becomes a character, and growing increasingly impatient with his protagonist, kills him because "he no longer needs him" for the broader story about imperialism he wants to tell (76). By now, the novel has turned into a collection of short dramatic scripts, retelling the latter stages of the Revolution as an abbreviated and irreverent pageant that includes a political "Puppet," an "Improvised *Caudillo*," a group of "Vultures," and "What Looks Like the People" (*Panchito Chapopote* 66–67). A "Commentating Chorus" humorously reports on the death of the protagonist: "thus ended the life of a nobody, who accomplished nothing as he was about to do something" (77–78).

As Icaza would later explain in *The Mexican Revolution and Literature*, the novel aims to offer a wider view of the political process (43–44). This is why even within the frame of Panchito's initial account, the narrator takes over and focuses on an event that could not have been witnessed or understood by the naïve protagonist. While Panchito is still dumbstruck by having sold his lands, the heads of the American and British contingents, still in search of more oil, sign up a decree partitioning a neighbouring ranch. As soon as they ink the deal,

> the Englishman becomes John Bull, and the American, Uncle Sam. The signing is a solemn act . . . John Bull thinks his son has grown up too much; he needs to be taught a lesson. Uncle Sam feels taller than his father. He believes he will take over his old lord.
>
> (*Panchito Chapopote* 50)

At this moment, "the Ruler of the waves lines up. The Yankee anthem. God save the King. Tipperary. Yankee Doodle . . . Uncle Sam and John Bull salute each other and stand at attention. When they say goodbye, they become business men again" (*Panchito Chapopote* 50–53). In Alva de la Canal's woodcut accompanying this passage, Uncle Sam and John Bull indeed salute each other and stand at attention. Between them, we see a table with a pen and a map of Mexico cut in half. Over them, a thunderstorm breaks and oil towers reach up to the sky. While Panchito can only understand the 'curse' on his lands in terms of agricultural infertility, and the corrupt government officials benefiting from the sale see no curse whatsoever, the novel itself alludes to a different kind of curse: the natural resources of a smaller nation attracting imperial interests. In this way, text and image work together towards an allegorical indictment of Anglo-American corporate greed, leaving behind the particular plight of the novel's protagonist.

Around the time that Icaza was writing these lines, Jara and Maples Arce were suing El Águila Oil Company, which was owned by the British oil company, Shell, and *Horizonte* was publishing articles explicitly condemning the labour and business practices of these companies.[14] In fact, *Irradiador* (*Irradiator*), a short-lived magazine published by Maples Arce and the estridentistas in Mexico City in autumn 1923, had already established the struggle for oil as a central political concern of the movement, coexisting with their avant-garde writing. *Irradiador* included two pieces by G.H. Martin on the Anglo-American rivalry over oil ("La rivalidad británico-americana y el petróleo"), and the first of these warned: "We are beginning to live in the age of oil, and the nation who controls the world's production will dominate the world" (*Irradiador* 16). These lines could very well serve as a leitmotif informing estridentista production during their time in Veracruz. In an article contained in the third issue of *Horizonte*, Marcelino Domingo denounces "the spectacle of the great political empires trampling over their own moral authority and universal credit in order to expand and impose their economic imperialism" (103). The piece then explains how at "the moment [oil] acquired these previously unsuspected applications," as fuel for battleships with the invention of the combustion engine, its exploitation "ceased to be a private and peaceful industry"; from that moment on "an empire's power was established by its ability to obtain and retain fossil fuels" (103). After giving a brief account of the Anglo-American rivalry across the world, Domingo condemns their disregard for the sovereignty of other nations:

> Neither has respected any rights in places where these rights are easily crushed; they have respected no sovereignty in places where sovereignty is easily dissolved; they have taken into account neither the limits established by their own laws nor the barriers put forth by foreign ones.
>
> (104)

Domingo concludes his piece by insisting that "the country that has incurred the greatest excesses is the United States, and the country that has suffered the most from these is Mexico" (104). Domingo's point here is not to measure the suffering of these colonies and former colonies, but to align *Estridentismo* with a point already established by the Mexican Constitution of 1917, which informed Jara's struggle against foreign companies, and which would eventually lead President Lázaro Cárdenas to nationalize the industry in 1938: Article 27 of the Constitution established the federal ownership of land and natural resources.

It was precisely this 'spectacle' of a ruthless economic imperialism that Icaza sought to give aesthetic form in *Panchito Chapopote*. In her reading, Flores conceives of Icaza and Alva de la Canal's collaborative project as an attempt to "negotiate the contradictions of addressing local matters in a modernist formal language" (261). By focusing primarily on Alva de la Canal's cubist-influenced woodcuts, Flores sees the local aspects limited to the level of content, while the vanguardist forms point to a cosmopolitan, and hence not strictly Mexican, aesthetic. She supports this interpretation with a compelling reading of one of the novel's most memorable scenes: after Panchito gets his nickname, the people of his hometown of Tepetate throw him a party. To celebrate, they decide to shoot down all the street lamps in town. Then, having no more light bulbs to destroy, and wishing to continue with their celebrations, they take aim at the moon: Panchito shoots it down, covering the whole sky in oil. At this very moment, a "drunken poet from France" responds: "bonsoir la Lime [sic; lune]!" (*Panchito Chapopote* 24). Since in *Actual No. 1*, Maples Arce had quoted the futurists, calling for the "murder of Clair de lune," Icaza here quotes both the European avant-garde as well as his estridentista predecessor. For Flores, this double-quote illustrates the

> effects of cultural imperialism. Just as Panchito becomes modern when his action is recognized by a French poet, Mexican artists are forced to look to Paris as their point of reference and must judge their work according to European standards.
>
> (261)

Flores here points out the connection between cultural and economic imperialism: as the moon bleeds oil, European intervention, aesthetic and political, covers everything in darkness. Indeed, Icaza wrote his novel after having spent several months in Europe, primarily visiting his friend, Mexican writer Alfonso Reyes, in Paris. In his letters, written just a month before composing *Panchito Chapopote*, Icaza expresses his disappointment at seeing Mexico in comparison to Europe (Zaïtzeff 145–146).

However, this is not the whole story. Icaza's use of vernacular speech and his incorporation of popular music, particularly *sones, rumbas, corridos, boleros*, and *huapangos*, structure the entire novel. Thus, it cannot be simply taken as an expression of local concerns articulated in European forms: instead, the novel itself becomes an instance of peripheral aesthetics. *Panchito Chapopote* both opens and ends with a *rumba*: an Afro-Cuban dance and musical genre that had been introduced to the Port of Veracruz by the early twentieth century (Enríquez Ureña 444). In the first scene of the novel, Porfiriata, "an old madman who sells newspapers and lottery tickets, and who thinks he's the reincarnation of heroes," begins to dance a *rumba* (*Panchito Chapopote* 8). Icaza describes the witnessing crowd as becoming a collective body: their breath "becomes a single gigantic mouth. The human mass gets confused in a single, oily and shaking [being]" (11). With this description, Icaza announces his focus on the region of Veracruz and the country of Mexico as collective bodies, rather than a concern with the fate of specific characters. In addition, Icaza interrupts the narrative with a back and forth between Porfiriata and the witnessing crowd that turns this part of the novel itself into a popular song. This dialogue between dancer and audience, in which Porfiriata's act of selling lottery tickets becomes the song's lyrics, puns on Panchito's frustrated sexual desire, after he has failed to marry his love interest in spite of his new wealth: "This ticket will get [the prize], this ticket will give it to you . . . / Your mouth is watering, *chico*!" (11). Moreover, Icaza uses popular humour to undermine the solemnity of the Anglo-American enterprise, to attack the corruption of the Mexican officials who profit from it, and to laugh at the impotence of poor, rural Mexicans like Panchito. As the military guard protecting the American contingent approaches Panchito's town, the narrator comments: "The gendarme's footsteps would have resonated if there had been pavement in Tepetate, and if he could afford shoes" (28). By the same token, a dialogue between two parrots mimics a conversation between the representatives of the American oil company, exposing the latter's ruthless plans and mocking the moral righteousness they had displayed earlier in the novel: "Lease land, any price, lease land, any price . . . Settle titles, fake titles, fake people" (42).

Nevertheless, beyond this recourse to popular humour and anti-imperialist irreverence, Icaza's irony conveys a sense of disillusionment with the Mexican Revolution and its leaders. In the prologue to his other major avant-garde work of the period, *Maganvoz, 1926*, Icaza concludes that a "farce" is the most appropriate genre to depict contemporary Mexico (16–17), as the epigraph at the beginning of this section suggests. Icaza's conception then not only explains *Panchito Chapopote*'s short, dramatic vignettes, but also how he represents the corruption of revolutionary leaders and the mediocrity of the masses. Tired of the armed conflict, the chorus labelled "What Looks Like the People" demands more bullfights, while "The Improvised *Caudillo*," speaking in sound bites, offers a caricature of the cause: "Fellow countrymen! Mexican people. The tyrants. The people. The vote. Vote suppression. The tyrants. The people. The suffrage. The imposition. I'll save the people. The people call for me. I sacrifice myself for the country. The vote, the tyrants" (67). Icaza's music lyrics, which he intersperses with the narrative, deflate these political leaders, especially in what we can label an "*anti-corrido*," which his narrator quotes during his account of the Revolution:

> I wish I could recount
> The feats of great men,
> Bestowing the fatherland with the honour
> Of their heroic deeds.
> But misfortune has made
> For ill-fated times
> In which even great exploits
> Are covered in mud.
>
> (54)

By associating mud with oil, the novel establishes a symbolic link between empire and the leaders of the Revolution, staining the entire political process.

The novel then extends this link further: towards the end, it provides a compressed summary of the Revolution, leading up to Álvaro Obregón's presidency (1920–1924), the first somewhat stable government after the armed conflict. First, by giving an account of the Revolution as a mere summary of violence, the novel crudely deflates its aims: "Life in the country is destroyed. Everyone suffers, all lose. Blood, suffering, and tears. The four tragic riders charge on implacably. Destruction. Ruins. Pain. Kidnappings. The forced loans continue. Bourgeois purses get emptied by force. Rebel militias are insatiable" (83–84). Then, once Obregón's presidency is established, the novel quickly points out how "Wall Street seizes the opportunity: it's convenient to be friends with the Government. Let's be friends with the Government. Let's help the Government. It foresees a lot of oil. Contracts. Big companies. Perks" (84). Icaza's critique of Obregón is important: in spite of the president's staunch support for the revolutionary project in certain areas, such as in public education, in 1923 he signed the Bucareli Treaty, which allowed American companies to continue to exploit Mexican oil. Moreover, Icaza's allusion to the federal government working in collusion with the US would turn out to be prescient, since President Calles's support of foreign oil companies would lead to the downfall of Jara's governorship.[15] After establishing his collaboration with the US, the novel quickly turns to Obregón's violent suppression of the right-wing rebellion led by his own finance minister, Adolfo de la Huerta, who in defiance of the Bucareli Treaty took up arms against Obregón's presidency (the US provided the Obregón regime with arms and planes that bombed de la Huerta's supporters). With dizzying speed, leading towards the end of the novel, Icaza depicts the outcome of this violent conflict as a "pyramid of human corpses," with Obregón as "the Commander in Chief" standing on top of them as though he were a "lion tamer" (89). Icaza then sets up a stark contrast between the crudity of this political violence and how the people of Veracruz experience it, "pleasurably dancing to the *rumba*" (89). In fact, by the end of the novel, Porfiriata's initial dance has become "symbolic and gigantic": not of the Revolution, but of a "popular" party (89, 93). With this, *Panchito Chapopote* ultimately portrays the Revolution and its immediate aftermath as a process that has not significantly transformed the lives of most Mexicans. At the very end of the novel, rural Veracruz goes back to its "old idyll" (94), thus suggesting a different meaning of "tropical," playing with its negative stereotype of a lazy and dissipated life.

How to reconcile this disenchanted ending with Icaza's lecture, where he insists on the novel's commitment to a socialist project he saw coming to fruition in the 1930s? On the one hand, Icaza's lecture responds to his aim to revise the politics of his literary production of the 1920s after he had experienced significant changes in his political outlook, participating, for instance, in the establishment of the *Universidad Obrera* (The Worker's University), as Rashkin observes (*Stridentist Movement* 216). On the other hand, in spite of depicting a chaotic world, *Panchito Chapopote* does not merely avoid commitment to the revolutionary project. More than an irony that refuses to be pinned down, the novel satirizes the Revolution and Obregón's presidency, and in doing so, denounces a process that was not living up to its potential. Here, it would be fruitful to return to Brennan's reconstruction of the Vichian tradition of anticolonial thought. In considering the hostility towards irony in this tradition, Brennan acknowledges that, in the guise of satire, irony is "recuperated somewhat": "a dissimulation is proper when it is necessary, but only if its saying the opposite can be understood clearly by readers as a calculated deviation from its native meaning" (Brennan 33). Icaza's novel repeats the aforementioned broken speech by the "The Improvised *Caudillo*," no longer as his initially hesitant words, but as the manifesto of an established revolutionary leader. This repetition, alongside the piles of corpses towards the end the novel, reveals how we are not supposed to take the "disinterested sacrifice" of these leaders literally (73). Moreover, in brief yet clear passages, the novel links political disillusionment to the collaboration of pre- and post-revolutionary governments with American oil companies:

[Pre-revolutionary dictator] Don Porfirio's government . . . is afraid that something might happen to [old Uncle Sam] in search for oil . . . Many years would have to elapse, and much blood would have to be spilled, before [Mexico] would learn to laugh at him.

(27)

Later, while recounting Obregón's repression, the novel mimics one of his speeches, and in doing so, condemns American interventionism: "The American government does not recognize the rebels. It is on the side of the Government. Legality must be upheld. It will not tolerate the free export of arms. It will only sell them to the Government" (85). In all of these instances, then, Icaza's satire of the Revolution can be read, in Brennan's words, as "the purloined letter, he conceals his meaning superficially, which is to say that he hides his meaning openly, enlisting a form of irony that joins the public in derision rather than playing behind its back" (34). Icaza hoped his readers would aim their derision at the corruption of the revolutionary project rather than at the Revolution itself.

Avant-garde commitment: *Horizonte*

A second moment in a Revolution . . . needs to justify the violence . . . and offer a better life than the one it destroyed.

(*Horizonte*)

Horizonte, published between April 1926 and May 1927 by the government of Veracruz, significantly moves away from the irony and opacity that characterized the movement's initial manifestos, as well as the writing that some of its members continued to publish at that time: most prominently List Arzubide's *El movimiento estridentista* (1926) and Arqueles Vela's *El café de nadie* (1926). *Horizonte* represents the emergence of a politically committed avant-garde that operated as a propaganda outlet for Jara's government, a compendium of nationalist culture and didacticism, and an instance of a formally diverse, peripheral aesthetics. This publication of "contemporary activity," as List Arzubide subtitled the magazine, combined more recognizably estridentista writing, such as poems by Maples Arce, List Arzubide, and Kyn Tanyia, with a vast array of editorials, articles, and reports on politics, history, rural and industrial economy and education, and popular culture. Among others, *Horizonte* published reports celebrating the construction of public works by the state government (Veracruz's first radio station and modern stadium, roads, housing, etc.); an article on Soviet public education by Lunatcharsky; another article on educational reform in Mexico by the then Under-Secretary of Public Education, Moisés Sáenz; articles on the oil and mining industries in Mexico; an essay on the Veracruz musical genre *danzón* by List Arzubide; another one on Mexican mural painting by Leopoldo Méndez; an article on taxation and agrarian reform by Leo Tolstoy; and many articles on Mexican historical figures from the Independence and the Revolution.

The bitter rivals of the estridentistas, the avant-garde poets and writers known as the *Contemporáneos*, were harshly critical of this estridentista collaboration, suggesting that this apparent marriage of mere economic and political convenience compromised their ethical and aesthetic integrity.[16] In response, the estridentistas defended themselves against these attacks, insisting that the state sponsorship they received was no cause for "shame" (*Horizonte* 199). Moreover, they did not hide their role as Jara's propagandists, especially in their reports on government public works. But even though the magazine has garnered a lot of interest recently, most scholarship continues to treat it either as an awkward compromise with the Veracruz government or a contradiction of their avant-garde aims. The reasoning behind these views is that an adversarial stance and the attack against the institution of art, which would exclusively define avant-garde production, must necessarily get co-opted when participating from within the institutions of the post-revolutionary regime and their nationalist ideology.[17] Even some of the same scholars who highlight *Estridentismo*'s combination of avant-garde aesthetics and political commitment end up considering this a "tension" that must be reconciled. A. Stanton, for instance, rightly reminds us that the existence of *Estridentismo* "belies the premise that a revolutionary movement in politics cannot be at the same time a revolutionary movement in aesthetics" (*Modernidad, vanguardia* 17). Indeed, becoming political actors was precisely how the *estridentistas* sought to join art with praxis. Yet, A. Stanton ultimately regards this alliance as generating "tensions," since the state will always seek to "control, use, and, if necessary, suppress the expressions that could threaten the new [political] order" (*Modernidad, vanguardia* 14). Ida Rodríguez Prampolini (1981) was one of the few scholars who approached this political phase of *Estridentismo* on its own terms, without seeking to reconcile it with Eurocentric conceptions of the avant-garde that oppose it to institutional collaboration (44).

One reason for considering *Horizonte* a contradiction is the deep-seated ambivalence in Mexican thought since the 1980s towards the post-revolutionary project and state, especially its construction of a 'national culture'. As Sánchez Prado points out, the last decades have been marked by a significant reassessment of the post-revolutionary project. In Sánchez Prado's reading, these reassessments have insisted on the "tension between the regional realities of a diverse national culture and an ideological and intellectual tradition that seeks to homogenize it with the aim to constitute or legitimize political power" (3). These essential critiques of the distortions and exclusions of the post-revolutionary project, particularly in its simultaneous creation of indigenous mythologies and the actual exclusion of indigenous groups, has led to a common sense that views with suspicion, if not downright hostility, *any* collaboration with the nationalist ideology and policies that emerged in that period, treating them as though they were an hegemonic monolith. Nevertheless, here it must be said that some of the very reassessments that Sánchez Prado considers, such as those carried out by Claudio Lomnitz (1992), do not simply discard the importance of envisioning a national, and even *nationalist*, project, given Mexico's precarious sovereignty. Even though Lomnitz rightly sets out to study 'national culture' in sociologically concrete terms, rather than merely reiterating the official ideology of the post-revolutionary state or even the insightful generalizations about 'national character' found in essays by prominent public intellectuals, such as Octavio Paz, he understands that the "problem of national culture" in Mexico has not been "abolished": the reason for this is that nationalism emerges precisely at critical moments when the "possibility of creating, shaping, and running one's own institutions" becomes threatened by foreign elites (14). It is certainly no coincidence that the distancing of Mexican intellectuals from the post-revolutionary project occurred during the consolidation of neoliberal thought, at the time when the Mexican economy began to open itself up to the global market. By the same token, and as Lomnitz presciently warned twenty-five years ago: "the fully antinationalist stance that has emerged as the backdrop of current official policy [and the crisis of Mexican nationalism] is unrealistic" (14). The reason for this is that, in the pre-NAFTA years, and even more so today, "to abandon all forms of nationalism is merely to place the country at the unqualified disposal of the market and of United States policy" (14). As Lomnitz elaborates, the main issue with many of these critiques of nationalist discourse is that they "do not fully address the available political alternatives" (14).

The estridentistas produced *Horizonte* right at another time of national crisis, when the fear of American intervention was well founded.[18] As Jara's troubled governorship reveals, the post-revolutionary project, or more accurately *projects*, was developed at the time when the American empire consolidated and the British Empire continued to make its presence felt throughout Latin America. In other words, more than the mere adoption of a theoretical position, *Estridentismo*'s nationalism entailed the assertion of sovereignty when its viability was not guaranteed, and which they articulated in aesthetic as much as political terms. Their aim was to envision a political alternative by embracing the most radical strands of the Revolution. Jara's administration followed several aspects of this project, particularly the interpretation of the Revolution articulated by José Vasconcelos, the founder of the Secretariat of Public Education in 1920, and its Secretary until 1924, at the end of Obregón's presidency. Vasconcelos, the first major ideologue of the post-revolutionary regime and the first major promoter of Muralism, had argued that the armed conflict would have to transition to the "battle fields" of culture and education, where writers and intellectuals like him would lead this new constructive phase (Blanco 83–84). This is why the estridentistas saw no contradiction between art making and an explicit political commitment, a view that informed Icaza's later conception of the avant-garde. But, distancing themselves from Vasconcelos' bookish education, the Education Under-Secretary under President Calles, Moisés Sáenz, and Jara's government put forth a materialist understanding of the role of culture and education, emphasizing how they should foster economic improvement, a view thoroughly developed in *Horizonte*. Still, all of them insisted on the importance of transitioning from the armed conflict to a cultural revolution. In *Horizonte*'s first issue, the unsigned propaganda piece for Jara's government quoted at the start of this section establishes two moments in a revolution: one of destruction of the past and one that needs to "justify the violence" and "offer a better form of life than the one it destroyed" (18). This "new hour of responsibility" would entail the construction of public works and institutions that would finally incorporate agricultural workers and the working classes into the new state (*Horizonte* 18–19).

Thus, *Horizonte* conceives of cultural practices as socially useful activities. A salient feature of the magazine is the relative absence of ironic writing and, by contrast, the strong presence of a didacticist ideology. In a column on rural education contained in the magazine's second issue of May 1926, the Director General of Education of the State of Veracruz, A. Pérez y Soto, calls on communities to build their own rural schools, in this way seeking to involve them in the creation of their new institutions (43–44). This piece, along with editorials on university reform, insists on the incorporation of useful technical and manual education into forms of non-instrumental knowledge. For Pérez y Soto, the reconciliation of these kinds of learning would result in a newfound "love of country" (44), a position that, needless to say, contrasts sharply with the movement's initial manifestos and even with most of estridentista literature still published during this period.

The magazine complements this emphasis on working-class and rural education with didactic literature. In the same issue of May 1926, *Horizonte* includes a short story by Ricardo Flores Magón, the early leader of the Revolution. His story, 'La catástrofe' ('The Catastrophe'), is a morality tale on mining labour and the price workers paid for not joining the Revolution. The third-person narrator tells this formally conventional story from the point of view of Pedro, a miner who has time to regret his refusal to join the armed conflict after his mine collapses and he gets trapped inside. The narrator recounts the protagonist's final thoughts by blurring his point of view with Pedro's through a set of indignant, rhetorical questions:

> Didn't they all remember at that precise moment that, in order to maximize their profit, the bourgeoisie had refused to provide them with sufficient wood to build the galleries inside the mine, and that the gallery in which the catastrophe had occurred was the worst off?
>
> (73)

The story closes with a lesson on the consequences of this refusal: the death of these miners, including the protagonist, and the crippling debt incurred by their poor families. To be sure, Flores Magon's narrative avoids an even more programmatic stance, for example by not specifying the revolutionary faction that another miner, Juan, decides to join: Juan's was not the cause of any particular party, but a "libertarian duty" against exploitation (72). By the same token, Pedro never learns of Juan's fate, and as he is about to die, merely imagines him on the battlefield fighting against his oppressors (74), in a scene reminiscent of the celebrations of male violence found in the novels of the Revolution, especially in *Los de abajo*. Still, in spite of avoiding allegiance to a specific faction, 'La catástrofe' lacks not only the humorous experimentation of *Panchito Chapopote* but also its ironic disillusionment: the story's commitment to the Revolution is emphatically clear and even unashamedly heavy-handed.

Nevertheless, this didactic story is only one kind of writing found in a magazine displaying, above all, the lack of a homogeneous aesthetic programme. Immediately following Flores Magon's story, the May 1926 issue includes the first canto of Maples Arce's 1924 poem 'Metropolis: Super-Bolshevik Poem in Five Cantos', accompanied by a reproduction of a cubist-inspired woodcut of an urban landscape by Alva de la Canal. The differences between these two texts are reinforced by the visual contrast between Alva de la Canal's woodcuts for 'La catástrofe', especially the first one, a figurative representation of two miners sitting under a tree with a mine in the background, and Alva de la Canal's urban landscape for 'Metropolis'. Maples Arce's poem follows some of the avant-garde patterns of his early estridentista poetry, written in free verse, seeking to create a montage effect by juxtaposing quasi-cinematic urban imagery, and strengthening this effect by laying out odd-number stanzas on the left side of the page, and even-number stanzas on the right. 'Metropolis' celebrates the "new, muscular beauty of the century" by expectedly likening an idealized worker's body to the steel constructions of an also idealized city (*Horizonte* 75–76). At some moments in the first canto, Maples Arce's tone becomes bathetic, as when he celebrates the "sexual fever / of the factories" (76). Still, this excitement for modern technology, along with a predictable catalogue of references to harsh "sounds," "motors," and "mechanical wings" (75), is consistent with his early estridentista poetry. The canto then focuses on large crowds of workers, who, marching along the metropolis, make "bourgeois thieves tremble" (77).

And yet, *Horizonte* tellingly omits the following cantos of the poem, where Maples Arce's enthusiasm for mass culture and political upheaval becomes more ambivalent and sombre. As Evodio Escalante (2002) argues in his insightful reading, "in spite of flirting with the triumphalism represented by the emergence of a working-class superhero," the poem displays a prominent "ambivalence and romantic nostalgia" that associates social revolution with destruction and desolation (49–50). According to Escalante, this ambivalence defines the (early) estridentista avant-garde, with one eye towards the future, celebrating technological progress and political revolution, and the other towards the past, lamenting the present violence and chaos. As Maples Arce would later explain in his autobiography, *Soberana juventud* (*Sovereign Youth*, 1967), he wrote these lines in 1923 during de la Huerta's rebellion against Obregón. Curiously, even though by the time the poem was published in 1924 the outcome had been positive for Obregón, Maples Arce's poem retains the anxiety he felt at the time. Having experienced dread and fascination as he watched the workers march on the streets, he wanted to capture his "feeling of hope and helplessness" at this sight (qtd. in Escalante 53). Thus, while the first canto closes with a faint sense of hope, with the poet wondering whether "the fiery embers of his lines / will shine on humiliated horizons" (*Horizonte* 78), the final canto closes at night with an image of an empty city burning in ruins: "The streets, / noisy and deserted, / are rivers of shadows / flowing toward the sea, / and the sky, unraveling, / is the new/ flag, / burning / over the city" (Schneider 434). Crucially, Maples Arce links this image of desolation to the destruction caused by political strife: the "unraveling sky" becomes a flag that hints at the labour movements of the period. For Escalante, this "unraveling sky" represents a "horizon that cannot promise anything" anymore and that "closes itself down," leaving the city in complete darkness (62).

By the time List Arzubide was in charge of estridentismo's government-sponsored magazine, the movement had charged this word, *horizonte*, with a decidedly hopeful connotation, associating it with the possibilities of Jara's political programme. It is no wonder then that Maples Arce's earlier sombreness and ironic equivocations no longer fit with the movement and their magazine. Yet, the estridentistas did not consider this distancing from their earlier aesthetic a mere abdication or co-option. As *Horizonte* makes clear, they considered that their avant-garde production during this constructive phase, however troubled and unstable, entailed a different outlook. Their collective art making now required their involvement in the creation of new institutions, rather than the attack against traditional aesthetics: it needed to veer off from destructive irony and propose a better life than the one the Revolution had destroyed.

Notes

1 Luis Mario Schneider's pioneering studies of the 1970s and 80s did much to revive interest in *Estridentismo* after decades of neglect. At that time, focus on the movement was predominantly concentrated in the State of Veracruz, where the estridentistas had produced much of their work. In 1981, Esther Hernández Palacios organized a symposium on the movement at the University of Veracruz, later published as *Estridentismo: memoria y valoración*, and during the same year, the journal of that university, *La Palabra y el Hombre*, dedicated a special issue to *Estridentismo*. However, while there were some groundbreaking studies of the movement in the 1990s, especially the catalogue to the 1991 exhibit in Mexico City, *Modernidad y modernización*, and Vicky Unruh's impressive study of Latin American avant-gardes (1994), it was not until the early 2000s that we saw a sustained effort to reclaim *Estridentismo*'s rightful place at the centre of post-revolutionary avant-gardes from the disciplines of art history and literary studies, both within the Mexican and American academies. See Escalante; Flores; Gallo; Klich, 'Estridentópolis' and *Revolution*; Pappe; Rashkin, 'Estridentópolis' and *Stridentist Movement*; A. Stanton, *Modernidad, vanguardia* and *Vanguardia en México, 1915–1940*. For a lucid account of the reasons for the estridentista exclusion in Mexican intellectual circles and the academy, see Escalante 9–40.
2 The term 'the novel of the Revolution' refers to the novels depicting the Mexican Revolution, written during or after the armed conflict. The most prominent example was Mariano Azuela's *Los de abajo* (*The Underdogs*, 1915), which the estridentistas helped canonize during the debates on national literature of 1925 and which they reissued in 1927.
3 As Claudio Lomnitz (2016) argues, the concept of "dependency" better explains the relationship between Latin America and the capitalist metropolis than the term "postcolonial." This is due to the region's "unusually long" history of national, independent states since the early nineteenth century, which highlights the "tension between capitalist development, modernization, and modernity" (*Nación desdibujada* 66). Once dependency established itself as a theory in the 1960s, it reinforced the already existing scepticism in the region regarding a "link between progress and national sovereignty," and it challenged capitalist notions that view "underdeveloped" nations as lagging behind developed ones (66). Instead, dependency theory "insisted on how development was not the fruit of underdevelopment, but its evil twin," and thus on how both were contemporaneous (66–67; my translation).

4   To be sure, Gamio, a student of Franz Boas, is best known for his project of '*mestizaje*', which has been intensively studied and criticized in recent decades, given its suggestion of an indigenous 'primitive' that must be incorporated into the modern nation. See Swarthout 95–105.

5   Recently, Anthony Stanton (2014) has rightly taken aim against stereotyped, and stubbornly resilient, accounts of post-revolutionary cultural production. As Stanton points out, these accounts cannot do justice to the various artistic groups of this period and the complexity of their aesthetic propositions: above all, Stanton singles out the poets known as the *Contemporáneos* (Jorge Cuesta, José Gorostiza, Jaime Torres Bodet, Salvador Novo, and Xavier Villaurrutia, among others), the estridentista writers themselves (Manuel Maples Arce, Germán List Arzubide, Salvador Gallardo, Kyn Taniya, and Arqueles Vela, among others), and the various visual artists that collaborated with either of these two groups, or with both, at different stages of their careers. According to Stanton, traditional approaches tend to view the *Contemporáneos* as "subjectivist" aesthetes, opposed to the politically committed muralists and novelists of the Revolution, when, as Stanton counters, the latter were not oblivious to formal experimentation and the former were not "insensitive to the revolutionary aim of presenting a reality that had been suppressed . . . for centuries" (16). Moreover, as Stanton elaborates, these accounts do not leave much room for the estridentistas, who sought to be revolutionary in both aesthetic and political terms. In this way, Stanton offers a very clear and helpful counter-narrative to those opposing political commitment to formal experimentation. Nevertheless, as his vague "revolutionary aim of presenting a suppressed reality" reveals, and as I discuss in the last section of this chapter, Stanton still understands the realm of aesthetics and of political collaboration as two distinct spheres that are necessarily in "tension" (14). By contrast, I argue that the estridentistas conceived of aesthetics and politics as aspects of a unified cultural practice (crucially, they shared this with the muralists and the novelists of the Revolution and not with the *Contemporáneos*). See A. Stanton, *Modernidad, vanguardia* 11–35 (my translation).

6   All translations of quotations in Spanish are mine.

7   See Cordero, in *Modernidad y modernización* 64; Flores 210, 219, 240–244; Sánchez Prado 52–59.

8   For a detailed account of the tensions between the federal government and Jara's administration (1924–1927), see Koth 240–251.

9   Xavier Icaza occupies an unusual place in *Estridentismo*. On the one hand, it is difficult to establish the precise degree of his involvement with the movement, in spite of the fact that he contributed to *Horizonte* and that the *Talleres Gráficos* published his first major work, *Magnavoz, 1926*, written after he moved to Xalapa, when he was presumably in close contact with the estridentistas. As Serge Zaïtzeff points out in his study of Icaza's correspondence, we have little information documenting his relationship with Maples Arce and the rest of the group, and List Arzubide does not mention him in either of the two editions of his fictional account of the movement, *El movimiento estridentista* (1926 and 1967). On the other hand, scholarship, both at the time and especially recently, has insisted on Icaza's centrality to *Estridentismo*. Thus, given that Icaza's work has been consistently read within the estridentista canon, including by contemporaneous critics, its aesthetic propositions have to be considered in relation to those of the larger movement. See Beals 265, 275; Brushwood, 'Bases' and *Narrative Innovation* 19–21; Escalante 92–104; Flores 256–264; Rashkin, *Stridentist Movement* 203–221; Ruth Stanton; Unruh 50–55.

10 Icaza's critique of the foreign oil industry in his novel is even more remarkable considering that, at that time, he was in fact working as a lawyer for the British-owned El Águila Oil Company. This meant that when Jara's administration sued El Águila, demanding "payment of taxes withheld by the oil companies" (Koth 247), Icaza and Maples Arce probably sat on opposite sides of the negotiating table, with the latter representing Jara as secretary of state and, at certain points during Jara's absence

while in Mexico City, as interim governor. Nevertheless, as Rashkin cautions, "Icaza's personal views . . . were not necessarily sympathetic to his employer," since in his letters he complained to his friend Alfonso Reyes of his work on behalf of the company (*Stridentist Movement* 206). More importantly, and as Rashkin also mentions, during the administration of President Lázaro Cárdenas, Icaza would be one of the Supreme Court justices to ratify the nationalization of the oil industry. But, above all, *Panchito Chapopote* itself is strong evidence of Icaza's emphatic criticism of his employer's actions. See Koth 244–249; Rashkin, *Stridentist Movement* 206.

11 Icaza published a significantly abridged translation into English of his lecture in the journal *Books Abroad* in 1937; Icaza titled this version 'Perspectives of Mexican Literature'. Since this English version omits his detailed discussion of revolutionary aesthetics included in his longer lecture in Spanish, I quote the latter below and provide my own translation into English.

12 For a more detailed discussion of Icaza's lecture, see Gerzso.

13 See Brushwood, 'Bases' 161, 166–170, and *Narrative Innovation* 19–21; Escalante 92–93, 97–104; Rashkin, *Stridentist Movement* 211–217.

14 See note 10.

15 Karl Koth points out how documents discovered in the US Embassy in Mexico City in 1993 reveal President Calvin Coolidge's intention to invade Mexico in 1926, after the passage of the federal Petroleum Law, "which gave the [oil] companies one year in which to present their applications for the confirmation of property rights" (246). As these documents show, President Calles learned of these plans for a potential invasion and "threatened the new US president . . . by revealing that he had given copies to every foreign ambassador in Mexico, and, in the event of war, their plans and their motives would become known throughout the world" (Koth 246). It was precisely then that Governor Jara "dug in his heels, demanding payment of taxes withheld by the oil companies" (Koth 247). According to Koth, Jara's demands led President Calles to give "exclusive rights" to the federal government to "intervene in labour conflicts in the petroleum and mining industries," thus contravening Jara's radically federalist interpretation of Article 27 of the Constitution (Koth 247). These conflicts ultimately motivated Calles to promote Jara's removal from office in October 1927, and Koth is in fact sympathetic to Calles, especially given US pressures at the time. For a favourable assessment of Jara's entire political career, see Lara Ponte.

16 See Escalante 28–32; note 5.

17 Sánchez Prado simply dismisses this phase of collaboration with the state, viewing it as a "direct contradiction" of their initial attack against cultural institutions, crystallized by distributing leaflets on the streets of Mexico City and Puebla (57). Flores, by contrast, celebrates the political impetus behind this phase of the movement. Nevertheless, she still characterizes it as a moment that "exposed the tensions and contradictions involved in defining an avant-garde when its members were also social actors and political agents" (210). Similarly, while Cordero emphasizes the "social and philosophical character" of Mexican avant-gardes, she ultimately views the estridentista collaboration with Jara as the moment when their "rebellious spirit crumbled" (*Modernidad y modernización* 53, 54).

18 See note 15.

Works cited

Beals, Carleton. *Mexican Maze*. J.D. Lippincott, 1931.

Becerra, Gabriela, editor. *Estridentismo: memoria y valoración*. Mexico City, Fondo de Cultura Económica, 1983.

Blanco, José Joaquin. *Se llamaba Vasconcelos: una evocación crítica*. Mexico City, Fondo de Cultura Económica, 1977.

Brennan, Timothy. *Borrowed Light: Vico, Hegel, and the Colonies.* Stanford UP, 2014.

Brushwood, John S. "Las bases del vanguardismo en Xavier Icaza." *Texto crítico*, vol. 8, no. 24–25, 1982, pp. 161–170.

—. *Narrative Innovation and Political Change in Mexico.* Peter Lang, 1989.

Enríquez Ureña, Pedro. *La utopía de América.* Caracas, Biblioteca Ayacucho, 1989.

Escalante, Evodio. *Elevación y caída del estridentismo.* Mexico City, Consejo Nacional para la Cultura y las Artes, 2002.

Flores, Tatiana. *Mexico's Revolutionary Avant-Gardes: From Estridentismo to ¡30–30!.* Yale UP, 2013.

Gallo, Rubén. *Mexican Modernity: The Avant-Garde and the Technological Revolution.* MIT Press, 2005.

Gerzso, Christian. "Icaza's Untimely Avant-Garde." *The Battersea Review*, no. 6, Fall 016, http://thebatterseareview.com/critical-prose/295-xavier-icaza-s-untimely-avantgarde. Accessed 19 Dec. 2017.

Horizonte. 1926–1927. Mexico City, Fondo de Cultura Económica, 2011.

Icaza, Xavier. *Magnavoz, 1926.* Xalapa, Talleres Gráficos del Gobierno de Veracruz, 1926.

—. *Panchito Chapopote: retablo tropical o relación de un extraordinario sucedido de la heróica Veracruz.* Mexico City, Aloma, 1961.

—. "Perspectives of Mexican Literature." *Books Abroad*, vol. 11, no. 2, Spring 1937, pp. 155–158.

—. *La Revolución Mexicana y la literatura.* Mexico City, Palacio de Bellas Artes, 1934.

Irradiador: revista de vanguardia, Sep. 1923.

Klich, Lynda. "Estridentópolis: Achieving a Post-Revolutionary Utopia in Jalapa." *The Journal of Decorative and Propaganda Arts*, no. 26, 2010, pp. 102–127.

—. *Revolution and Utopia: Estridentismo and the Visual Arts, 1921–1927.* Dissertation, New York University, 2008.

Koth, Karl B. *Waking the Dictator: Veracruz, the Struggle for Federalism, and the Mexican Revolution, 1870–1927.* U of Calgary P, 2002.

La Palabra y el Hombre, Oct.–Dec. 1981.

Lara Ponte, Rodolfo. *Heriberto Jara, vigencia de un ideal.* Mexico City, Fondo de Cultura Económica, 2000.

Lomnitz, Claudio. *Exits from the Labyrinth: Culture and Ideology in the Mexican National Space.* U of California P, 1992.

—. *La nación desdibujada: México en trece ensayos.* Mexico City, Editorial Malpaso, 2016.

Maples Arce, Manuel. *Soberana juventud.* Madrid, Editorial Plenitud, 1967.

Modernidad y modernización en el arte mexicano, 1920–1960. Mexico City, Museo Nacional de Arte, 1991.

Pappe, Silvia. *Estridentópolis: urbanización y montaje.* Mexico City, Universidad Autónoma Metropolitana, 2006.

Rashkin, Elissa J. "Estridentópolis: Public Life of the Avant-Garde in Veracruz, 1925–1927." *Studies in Latin American Popular Culture*, no. 25, 2006, pp. 73–94.

—. *The Stridentist Movement in Mexico: The Avant-Garde and Cultural Change in the 1920s.* Lexington Books, 2009.

Rodríguez Prampolini, Ida. "Ramón Alva de la Canal." *Plural*, vol. 11, no. 123, Dec. 1981, pp. 41–45.

Sánchez Prado, Ignacio. *Naciones intelectuales: Las fundaciones de la modernidad literaria mexicana (1917–1959).* Purdue UP, 2009.

Schneider, Luis Mario. *El estridentismo o una literatura de la estrategia.* Mexico City, Consejo Nacional para la Cultura y las Artes, 1997.

Stanton, Anthony, ed. *Modernidad, vanguardia y revolución en la poesía mexicana (1919–1930).* Mexico City, Colegio de México, 2014.

— and Renato González Mello, eds. *Vanguardia en México, 1915–1940.* Mexico City, Museo Nacional de Arte, 2013.

Stanton, Ruth. "Development of Icaza as Leader in the Estridentista School of Mexican Literature." *Hispania*, vol. 21, no. 4, Dec. 1938, pp. 271–280.

Swarthout, Kelley. "Assimilating the Primitive." *Parallel Dialogues on Racial Miscegenation in Revolutionary Mexico.* Peter Lang, 2004, pp. 95–105.

Unruh, Vicky. *Latin American Vanguards: The Art of Contentious Encounters.* U of California P, 1994.

Zaïtzeff, Serge. *Xavier Icaza y sus contemporáneos epistolarios.* Xalapa, Universidad Veracruzana, 1995.

Part III
Poetic history

8 Vichian language and the Irish Troubles

Brian Friel's *Translations*

Sreya Chatterjee

I

Vichian materialism offers a *humanist* framework through which the two aspects of language, one as aesthetic self-expression and the other as political self-determination, can be articulated in their interrelatedness. This essay explores Vichian themes in political theatre. I examine Brian Friel's seminal play, *Translations* (1980), to trace the relationship between Irish theatre and the language of Irish anticolonialism. Language has played a key role in anticolonial movements in the European and global peripheries, yet much remains to be said about the relation of language to material struggles in the context of anticolonialism.

In *Borrowed Light: Vico, Hegel, and the Colonies* (2014), Timothy Brennan charts the materiality of language as elucidated by the Neapolitan philologist Giambattista Vico (1668–1744). Published in 1725, Vico's *New Science*, especially the section on 'Poetic Logic', foregrounds the importance of everyday, quotidian forms of social exchange as a historical record of human struggle and the evolution of civic institutions. Vico's twin insights, as emphasized by Brennan, are especially important for my essay: first, Vico holds "that specific ideas, linguistic innovations and forms of art correspond to a period's conditions of social organization" (20); second, he argues that human history is "prophetically non-denominational," which is to say that no one people, race, or ethnicity has priority over others. In charting a "gentile" human history against the grain of dominant historiography, Vico specifically dismisses, as Brennan notes, "the providential story of the chosen people," instead focusing on plebeian cultural and linguistic expression as political form (21).

Vico's philological project focused on the historical use of language is nothing less than a Herculean venture. It elevates oral linguistic expression as sites of human labour and striving. It is Vico, more than Spinoza and prior to Marx, who re-inscribes the human subject within materialist philosophy as an agent of historical transformation. As Brennan argues, Vico extols the human being, "without triumphalizing 'reason' as it was passed down from the scientific Enlightenment" (19). Brennan's *Borrowed Light* makes a crucial case for Vico and Vichian humanism as a forerunner of twentieth-century European anti-imperialism.[1] In his rejection of foreign domination as a form of class oppression, Vico anticipates some of the most powerful progressive currents of the twentieth century in Asia and Africa, and crucially, for my essay, *within* Europe.

The interwar era (1918–1939) witnessed the unprecedented triumph of sub-jugated peoples under the British, Tsarist, and Austro-Hungarian empires, at the peripheries and semi-peripheries of Continental Europe. This is generally a point that remains unacknowledged in standard accounts of postcolonial theory.[2] European anti-imperialist thought engaged with the non-European colonial world as well as with the movements for decolonization, the end of class and national oppression, and foreign domination. In the context of the British Isles, the formation of the Irish Free State in 1922 on the back of the fraught struggle of the socialist republicans and Sinn Féin heralded a critical articulation of national sovereignty. The early years of political independence in the Irish Free State were marked by a surge of idealism; this progressive nationalist vision translated into concrete achievements in social welfare, the establishment of constitutional safeguards, rural development, and progressive measures for the rights of women.[3]

In Ireland, republican socialist nationalism was gradually co-opted over the decades of the 1940s and 50s, as the Fianna Fail government of Éamon de Valera embraced social and economic conservatism over its earlier radical positions. Beginning in the 1960s, the Lemass government reverted the anti-capitalist stance of earlier republicanism in favour of accelerated Western-style development and a willingness to join the European Economic Community (EEC), a forerunner of the latter-day European Union. In Northern Ireland, the Protestant Unionists echoed similar claims of assimilating to *laissez-faire* Europe. On the other side of the political equation, the Northern Irish Troubles of the late 1960s directly illustrated the disenfranchisement and segregation of the Catholic minority. The 'postcolonial' trajectories of both the Republic of Ireland as well as Northern Ireland thus came to be marked by the consolidation of a majoritarian cultural and religious nationalism combined with the economic marginalization of minorities, women, and the working and rural poor.

The site of language highlighted this crisis of nationalism. In this regard, the Irish situation is usefully compared to the postcolonial peripheries outside of Europe, such as those in Africa and Asia, where visible disjuncture emerged between elite nationalist culture on one hand, and popular-front expressions shaped by distinct kinds of national belonging on the other. In post-independence Kenya, writes the distinguished Marxist intellectual Ngũgĩ wa Thiongo, national-ist culture promoted the colonial languages of English and French as the medium of middle-class self-fashioning. This meant the relegation of local, vernacular languages: the latter were only refurbished, wa Thiongo argues, by the diverse sections of the Kenyan underclass: "all these languages were kept alive in the daily speech, in the ceremonies, in political struggles, above all in the rich store of orature–proverbs, stories, poems and riddles" (*Decolonising* 23). Similarly, in postcolonial India, there was a resurgence of interest in popular theatre that was pioneered by the likes of Bijan Bhattacharya and Utpal Dutta, who refashioned traditional "folk" forms for the purpose of new, postcolonial critique.[4]

Thus the nation, the proper and legitimate horizon of decolonization struggles in Kenya and India as much as in Ireland, no longer aligned with the popular in the postcolonial conjuncture. By the second half of the twentieth century, the popular—that is, the cultural—practice of the broad masses had to situate itself *in opposition to* the national. This is because the revival of a cohesive and homogeneous national culture, promoted by the elite with myriad motives and agendas, stemmed from a regressive form of cultural chauvinism and aided in the consolidation of a majoritarian nationalism.[5] The oppositional impact of Vichian materialism reasserted itself at this conjuncture as was illustrated by the case of theatre in particular: a truly Vichian, sovereign popular theatre animated by what Brennan calls "the plebeian logic of the literary vulgate" (74). Such a theatre negotiated, and dialectically negated, the longing for a national form and addressed itself to a world-historical notion of decolonization.[6]

The case of Ireland exemplifies a historically specific relationship between nationalism and popular theatre. In Ireland, to an even greater degree than the ex-colonial societies of Africa and India, the English language was adopted into the rubric of national self-formation. The national theatres in Ireland used English as the dominant medium, in part to obfuscate the asymmetries of underdeveloped Irish society. Ideologically aligned with the conservative project of nostalgia, Irish national theatre either consciously excluded the more militant forms of Irish rural theatre, or deliberately reconfigured rural Ireland as an idyllic and decidedly apolitical trope.[7]

Such paradoxes were directly related to the anomalous position of Ireland as a periphery within the geographical and cultural West, as distinct from Asian or African postcolonial nation-states. Ireland's proximity to the British metropole shaped its cultural and political structures while sustaining its economic back-wardness, even in the postcolonial period. As previously mentioned, the Republic of Ireland pursued a policy of accelerated economic development and rapid indus-trialization beginning in the 1960s while Northern Ireland experienced continuous political instability due to the Troubles, an intense bout of sectarian strife between the Catholic working-class populations and the occupying colonial English mili-tary. Such contradictions in Ireland's recent history as the poster-child of progress as well as old-style colonial repression vexed its attempts to join the front ranks of the 'free' West.[8]

When the Irish dramatist Brian Friel wrote *Translations*, it was specifically meant to address Ireland's complex colonial history. The play, first staged in 1980 but set in the year 1833, anticipates the turbulent era of the Troubles and reflects back upon it. By emphasizing the simultaneous presence of the Troubles in the future as well as the past, the play formally replicates the combined and uneven space-time of Ireland.[9] The peculiarities of Ireland's social and political structures are underlined, equally powerfully, by Friel's use of English to rep-resent a rural, Irish-speaking community in the play. This strategic adaptation of English, the language of Irish national theatre, to depict a rural social milieu largely sidestepped by the latter brings out the contradiction of postcolonial modernity in the Irish context.

Furthermore, the evocation and centrality of the Troubles in *Translations* bear more than a passing imprint of Friel's background as a Northern Irish playwright. Friel's experiences growing up were divided between Derry, the seat of the historical Troubles, and county Donegal, across the border in the Irish Free State, where the play is set. The play, *Translations*, traverses a borderland that is as much political as it is geographical. While Northern Ireland is often associated with a political vision foredoomed by working-class militancy and sectarian conflict, in *Translations*, Friel re-imagines the Troubles as a condition of new possibilities that might shape a radically transformed Northern Irish society.[10]

Finally, with the staging of *Translations* in Derry in 1980, Friel introduced a new form of theatre to Ireland. The play is the first theatrical production of the Field Day Theatre Company, a cultural and intellectual movement launched by Friel and several others to specifically address the Northern Irish situation, and its implications for Irish culture as a whole. Previously, Irish nationalist theatre had flourished under the auspices of Lady Gregory, W.B. Yeats, and J.M. Synge in the early decades of the twentieth century. These dramatists often idealized less visible sections of the Irish periphery—peasants, landless labourers, and women—in their search for an authentic national form. The Field Day Theatre Company approached similar subjects, but with the distinct aim to revive a form of popular theatre that would address itself to the fractured, contemporary Irish reality after the Troubles.[11]

II

The play, *Translations*, articulates the complex history of Ireland at the level of form. Using elements of Irish national as well as folk theatre, it charts the advent of colonialism into rural Ireland in the 1830s. A series of miscommunications between British officials and a group of rural Irish characters leads to the outbreak of colonial repression and counter-resistance. Despite this violent denouement, *Translations* resists the tragic form so common to Northern Irish drama.[12] Instead, it uses a Vichian materialist notion of language to open the entropic form of Irish drama to the humanist possibilities of social transformation. Owen, one of the characters in the play who is assigned the role of translator, leaves the stage with the grim warning of imminent "troubles." In this way, the play's depiction of nineteenth-century anticolonial struggle is linked to the recent Troubles of the 1960s.

The play portrays the plebeian setting of a small farming community in the hamlet of Baile Beag (Ballybeg) in County Donegal. Instead of the pastoral landscape familiar to national theatre, the play highlights the poverty and crisis of the rural-colonial periphery. In doing so, it emphasizes the coeval-ness of peripheral culture as distinct from the national. The opening scene shows broken and dysfunctional farming tools stacked up on the stage, denoting a barn. The barn provides the main setting of the play: it serves as the schoolroom as well as the living quarters for the school's headmaster, Hugh, and his elder son, Manus. The overuse of the decrepit barn as the only available form of built space illuminates the harsh material realities of rural Ireland. It provides a necessary prelude to the social relations represented in the play.

The physical setting of the play is an obscure, poor hamlet. As well, its temporal setting in the year 1833 makes the play appear far removed from the context of its urban-based audience in the 1980s. Similarly, the rural characters' repeated use of Greek and Latin creates a sense of cultural disjuncture: the epic-heroic universe of classical mythology is in sharp juxtaposition to the humble world they inhabit. These elements set the stage for a new kind of theatre; things at once remote and exotic are brought to life in the play, not only as a commentary on the contemporary present, but also to help imagine a historical lineage of the *future*.

The acute poverty made evident at the beginning of the play accentuates the spirit of survival demonstrated by the characters. The first scene depicts a schoolroom: a place of learning and self-improvement. The makeshift schoolroom, ingeniously put together using discarded bench seats and stools from the barn, provides a place for many of the local residents to learn languages. The lessons are conducted by Manus who, at the opening of the play, is seen helping one of his students utter her name. Manus' effort is emblematic, in many ways, of the struggle for social progress: the play projects the characters as agents of social change, not figures of passive suffering.

Despite its positive message, the opening scene does not attempt to gloss over the material realities of the rural periphery. In fact, it foregrounds a Vichian notion of plebeian, working-class culture as the font of political self-expression. The students, a young girl named Sarah and an elderly man called Jimmy Jack Cassie, embody the 'pathetic' image of the rural poor. These characters testify to the pathological effects of structural underdevelopment. Sarah, who has a severe speech defect, is "waiflike" and "could be any age from seventeen to thirty-five," while Jimmy is an elderly tramp. Even Manus, their teacher, is physically disabled.

It should be noted that the childlike, innocent Irish peasant has been a common trope in Irish literature and drama, especially in national theatre. Supposedly embodying the authentic and idyllic beauty of rural Ireland, the figure of the innocent peasant is a largely metropolitan fabrication, fashioned in Dublin for Dubliners to bring about social cohesion and foreground a national identity (O'Toole 112). In *Translations*, by contrast, Friel presents rural characters who reinstate the trope of the poor and provincial Irish peasant, but who also comment on their own peculiar condition as the effect of an underdeveloped social milieu. This is both distinct from and critical of the romantic portrayal of rural Ireland in twentieth-century national theatre. Friel's revival of the Irish peasant as a figure of resistance rather than passive suffering underscores a progressive, liberationist social imaginary.

Translations juxtaposes the country and the city to better emphasize the paradox of combined and uneven development. Owen, Manus' younger brother, embodies the aspirations of the migratory petit bourgeois. Whereas Manus assisted their father Hugh in running the hedge-school, Owen moved to the city of Dublin. He is a successful shop owner in the city, and the proud possessor of "twelve horses and six servants"; in Baile Beag, on the other hand, "nothing's changed" since he left six years ago (27). The differential trajectories of the brothers Manus and Owen highlight social fissures, and especially illuminate the contradictions in Owen's processes of "becoming," namely his smug assimilation as a Dubliner. Owen draws his identity from a colonial city that, like him, has prospered at the cost of rural underdevelopment and impoverishment.

Owen's paradoxical role as the underling of colonialism is confirmed when he brings a group of British engineers with him to Baile Beag, his native village. As a mercenary and native informant, Owen is part of a Land Ordnance Survey team that intends to re-map Baile Beag for taxation purposes. The chief undertakings of this group are to measure the land, and to Anglicize the Irish place names. However, the local people of Baile Beag are already aware of their arrival. Hugh, the headmaster, meets Captain Lancey, one of the colonial officers. Captain Lancey expresses concern to Hugh about his horses and equipment going astray, a situation further exacerbated by the fact that the locals do not speak his language, English. The episode signals the impending crisis of colonial integration; at the same time, it indicates that the Ordinance Survey team are already facing minor acts of resistance from the locals.

The mystery of the vanishing equipment and horses is solved when two other characters, Doalty and Bridget, appear in the play. Doalty reports to Manus that the "red coats" were seen over the countryside, "dragging them aul chains and peeping through the big machine they lug about everywhere with them" (17). This "big machine," a theodolite, is an intricate contraption used in the measurement of land. However, it is also a symbol of the colonizer's 'othering' gaze. Filtered through the theodolite, this voyeuristic gaze acquires the power to feminize, control, and manipulate the countryside of the colonial periphery.

To Doalty's unsophisticated eye, the modern, scientific theodolite acquires the strangeness of an unknown beast. It appears almost divine: "*Theo—theos—* something to do with a god. Maybe *thea*—a goddess" (17). This divine emblem of colonial modernity is no match, however, for Doalty's ingenious native "magic," which he uses to confound and confuse the British engineers. Every time the engineers position the theodolite to take measurements of the land, Doalty creeps up behind the machine and shifts it a few paces, such that the calculations are rendered erroneous. Such an act, undertaken out of a spirit of jesting, nonetheless constitutes one of the many instances of local resistance to the activities of the imperial engineers. Tellingly, it pits the "magic" of popular revolt against the imposition of seemingly omniscient science and colonial rule.

The presence of the British generates a sense of grim foreboding, one that is augmented by rumours of the potato famine. Early on in the play, Doalty's companion, Bridget, talks about rumours of a "sweet smell," especially marked in the areas where the British are seen with their machine. The smell could be that of rotting crops, indicating the possibility of an impending year of bad harvest. Such a possibility is immediately dismissed by Maire Chatach, Manus' fiancée, who reminds the others that this rumour circulates every year around the same time, but never culminates in an actual disaster.

Even though Maire declares the rumours of the famine to be baseless, to the best of her understanding, the possibility of famine looms large over the play. The historical reference is to the Irish famine of 1845. In the play, the intertwining of the famine and the colonial act of land measurement provides a prolepsis of later events, when the invading British army physically destroys the harvest and coerces the people of Ballybeg into submission. While the earlier scenes of the play sustain a comic tone (and native naiveté) to describe the British "red coats," the reference to the famine indicates the grim outcome of the British incursion into rural Ireland.

The play, significantly, reconfigures the social function of language. *Translations* evokes multiple languages: Greek and Latin, English and Irish, which are simultaneously in use in the play's milieu. This simultaneity of vernacular, classical, and colonial languages in the play emphasizes the unevenness of the Irish periphery. The headmaster, Hugh, insists upon Greek and Latin as the preferred medium of instruction at the hedge-school. He even tells the colonial officer, Captain Lancey, that Greek and Latin are more suited to the Irish way of life than English, which is mainly suited for trade and commerce. At the same time, there are reports that a new English-medium national school is to open in the neighbouring town. Both Hugh and Manus vie for a position there, in the hope of making more money. This last bit is particularly suggestive: it indicates a rapidly evolving social reality where the knowledge of English, as well as the desire to "profess" English, is concomitant with socioeconomic survival.

Maire Chatach, Manus' fiancée, emphasizes the hegemonic position of the English language as a means of upward social mobility. She quotes a Kerry politician who has remarked that "the old language is a barrier to modern progress"; and she insists upon learning English, rather than Greek and Latin: "I don't want Greek. I don't want Latin. I want English" (25). Maire's plea, directed to Hugh, underscores both the panic and the powerlessness of a predominantly Irish-speaking community in the face of historic transformations. At the same time, it reveals the tragic recalcitrance of characters like Hugh, whose ideas about language are entirely out-of-sync with the changed reality of the colonial periphery. By repeatedly focusing on the characters' perception of language, the play opens up the ideological and material contestations over linguistic colonization for scrutiny.

The 'translation scene' involving Owen emphasizes the power of language to bring about an alternative reality. During the translation, undertaken by Owen and one of the colonial officers, Yolland, the latter expresses concern that this is "an eviction of sorts," and that "something is being eroded" as a result (43). The utterance of these words is followed by Owen's recitation of the Anglicized place names:

> Lis na Muc, the Fort of the Pigs, has become Swinefort . . . And to get to Swinefort you pass through Greencastle and Fair Head and Strandhill and Gort and Whiteplains. And the new school isn't at Poll na gCaorach—it's at Sheepsrock.
>
> (42)

The exchange contextualizes the erosion of the existing cultural coordinates of Ballybeg. Owen's recitation reveals, like the power of a chant, a 'colonized' landscape from which every vestige of history and memory is being removed.

As the powerless and passive recipient of this information, the elderly Hugh reacts sentimentally, and points out to Yolland the peculiar structural dependencies that are hoisted on language in the periphery. In peripheral underdeveloped societies, he says, popular language expresses a way of life, and especially a means of articulating everyday survival:

> You'll find, sir, that certain cultures expend on their vocabularies and syntax acquisitive energies and ostentations entirely lacking in their material lives . . . it is our response to mud cabins and a diet of potatoes; our only method of replying to . . . inevitabilities.
>
> (42)

Once the Irish language is lost, however, Hugh's words sound overly melodramatic: sentimental, precisely because they are powerless. They cause only momentary discomfort, but cannot alter an already colonized reality with which they have no correspondence.

The two officers, Lancey and Yolland, exemplify very different archetypes of the colonial officer. Lancey is the prototype of the zealous and determined imperial servant who has unwavering faith in the civilizing mission of the empire. Yolland, on the other hand, is a caricature of the benign administrator; he is enamoured of the exotic, pastoral beauty of Ballybeg and of the local Irish women. Unlike Lancey, the older, more seasoned officer, Yolland is learning on the job. In fact, he turns up in Ballybeg because he missed his boat to another colonial periphery, the port city of Bombay in India. Yet, Yolland's stupidity only underscores his destructive presence in the play. He seduces Maire, and then disappears without a trace, putting the lives of the local people in dire crisis.

Yolland's disappearance intensifies the already exacerbated relations between the British and the local residents. Manus, who had threatened Yolland after the discovery of Yolland's affair with Maire, is suspected of the crime. Manus decides to leave for Mayo to seek refuge with some distant relatives. In the meantime, Doalty and Bridget bring the ominous news of the army closing in on Ballybeg, wreaking havoc on the crops in their search for the missing officer Yolland. The sense of crisis hanging over the action from the very first scene now culminates in a tense exchange between Captain Lancey and the locals, who, perhaps for the first time, fully realize the danger posed by the colonial army and its overwhelming military might.

Captain Lancey wastes no time in announcing his intentions to the inhabitants of Ballybeg. Compelling Owen to translate his warning, he threatens, first, to "shoot all livestock in Ballybeg"; then to "embark on a series of evictions and levelling of every abode" in selected areas; and finally, to "proceed until a complete clearance is made of this entire section" (62). The fate that Lancey pronounces on the residents of Ballybeg thus far exceeds the boundaries of justice. Under the guise of searching for Yolland, Lancey sets out to achieve precisely that which he had been sent to do in the first place—the complete annihilation of a whole community. The genocidal overtones in Lancey's threat may be read, in fact, as an explication of the very process of colonialism. The killing of the livestock indicates depletion of accumulated wealth and resources; the levelling of houses, the physical demolition of built space; and the final words, "complete clearance," evoke the erasure of the entire community.

Yet again, the play elucidates the importance of language in generating resistance among oppressed, plebeian peoples. Lancey's words are met, almost instantaneously, with the outbreak of violence in the background. Even as Lancey delivers his grim warning, Doalty announces that several British army camps are on fire, possibly an act of retaliation by the elusive Donnelly twins, militant organizers against the British. Lancey rushes out, cursing and threatening the others with dire consequences. It is at this point that the long-suffering translator, Owen, declares his intention to join the Donnelly twins in resisting the British. He leaves the stage with a cryptic warning that resounds with his departing footsteps: "there may be trouble" (66).

Significantly, *Translations* does not end with Owen's transformation from collaborator and translator to anti-British insurgent. The final scene again turns to Owen's father, the headmaster Hugh, to provide a different kind of translation and closure to the action. Hugh evokes an epic-poetical conception of the final events, as is his wont in grandiose prose. The takeover of Ballybeg by the British reminds him of a similar episode in Virgil's *Aeneid*: "kings of broad realms and proud in war who would come forth for Lybia's downfall . . . such was the course ordained . . . by fate" (68). Ballybeg's downfall, which parallels the obliteration of the Irish language, must forge a new mode of survival stemming from a strategic adaptation of a new language: "We must learn where we live. We must learn to make them our own. We must make them our new home" (66).[13]

III

The ending of *Translations* with the news of the advancing army, and Hugh's lament about the destruction of language, has prompted scholars to read *Translations* as an ambiguous play about the loss of identity. "The consensual interpretation," according to Shaun Richards, "is one which reads Friel, through Hugh, as arguing for accommodating rather than resisting cultural change" (57). However, Hugh's final words suggest the continuity of Irish rather than its obliteration. The old language could survive, Hugh suggests, if the people find a way to survive within the changed geopolitical landscape of colonialism. The play's message of fateful reconciliation is not merely a gentle acceptance of defeat, as some have suggested, but a means to survive in a reconfigured context defined by persistent imperial violence.

I have argued that *Translations* foregrounds the Vichian humanist struggle over language, as well as the struggle within language: to be immediately relevant, to fulfil its specific function of collective self-expression, and to resist becoming artificial. Hugh's lament for the loss of the Irish language both conveys this impulse and contextualizes it within the specific framework of Irish national history. It is worth noting that the play, which is written in English, reiterates the relegation of the Irish language in the colonial and postcolonial periphery. At the same time, its adoption of English is anti-parochial and non-nativist, and does not discount the potential of the latter language to be politically combative.[14]

The ending of *Translations* points to a paradox. The play ends with Hugh's lament that follows after Owen's mention, prior to this scene, of violence in the countryside. Owen's statement on the possibility of anticolonial counter-violence of the oppressed keeps the text open-ended and inconclusive, while Hugh seems to be calling for closure. The play foregrounds its thematic preoccupation with language precisely through this tension; there is after all no realized portrayal, or action in the literal sense, of Owen, Doalty, and the Donnelly brothers resisting the British. The audience must rely on Owen's words alone to imagine the possibility of resistance from this latter group. Owen's words thus provide a parallel template to Hugh's emphasis on survival: in the absence of action, it is language that carries forward the hope of resistance.

Translations encapsulates a key Vichian notion, that of the materiality of language, in its emphases on the spoken word and the performative context of theatre. At the same time, it accords pride of place to what Vico would call 'poetic logic', namely the ability of literature to provide humanistic, non-instrumental knowledge of the evolution of social relations. The play does this through its reconfiguration of tragedy. Unlike much of Irish nationalist theatre, *Translations* does not use the form of tragedy to mount a nostalgic attachment to a lost social order, or a conservative longing for national form. The impulse of social transformation and change is inherent to the play, which begins by establishing the importance of education, and with characters that have either achieved a degree of cultivation and mobility, or try to do so. The interrogative and dynamic restlessness of the characters supersedes the structural and material handicaps in which they find themselves implicated. The impulse of change both heightens and gives new meaning to the tragic ending of the play.

Translations exemplifies an anti-conservative understanding of tragedy, seeing it as a form that articulates the birth of a new social order. In this, it is similar to the German tragic drama of Lessing, Goethe, and Schiller (a Vichian tradition, as Erich Auerbach demonstrates).[15] The impact of colonialism and neocolonialism, while undoubtedly catastrophic and traumatic, makes any return to the past impossible. The historical past can, however, serve as a resource and a lesson for the future. *Translations* uses tragedy as entropic form: marking the passing of a social order while releasing the necessary energy for myriad other possibilities. This entropy does not provide a blueprint or prescription for the future, except as a vision of that which is not available in the present. It is a *negative* vision, in other words, foreshadowing yet-to-be-actualized "troubles."

In the parlance of Vico's *New Science*, such poetic knowledge is crucial for determining the scope of human agency. *Translations* overcomes, at least at the level of poetic logic, the twin defeats of both the distant and recent pasts in Ireland. The former is marked as much by the Union Act of 1800 and the Famine of 1845 as the latter is by the Northern Irish Troubles of the 1960s and Irish attempts to join the West since the 1970s. *Translations* alerts us to the continuing legacy of emancipatory struggles within Continental Europe, at a moment when anticolonialism of the early twentieth century seems to be co-opted by nationalism and the capitalist world-system. What the play foregrounds is the range of human activity in history, and its relevance for understanding the evolution of the future. Language itself is understood as the richest site of human interaction and self-expression, in sharp contrast to those schools of thought that would hold language to be arbitrary, discursive, and constructing the human as a 'figure' rather than the other way around. The Irish anti-imperialist tradition that the play foregrounds thus expands our understanding of the humanist resistance to new forms of domination and exploitation.

Notes

1 Brennan breaks new ground in Vico studies, departing from the pioneering frameworks set earlier by Cassirer in German, and Pompa in English.
2 Brennan addresses this in an earlier work, *Wars of Position*, especially Chapter 1.
3 See Brown, *Ireland: A Social and Cultural History*.
4 Thiong'o, *Globalectics*, provides a recent formulation. For India see Dutt, *Towards a Revolutionary Theatre*.
5 The Subaltern Studies Collective attempted a similar critique of nationalism, incorporating the struggles of oppressed groups in the periphery, such as peasants and women. However, it eventually moved away from a class-based analysis to a discourse-oriented one, thereby relegating the question of class struggle.
6 I take this concept from Brennan's "National Longing for Form."
7 Pilkington provides a summary discussion, using examples of plays such as Gregory's *Spreading the News* (1904) and Synge's *The Playboy of the Western World* (1907).
8 Irish postcolonial scholars are particularly attentive to this dilemma. Gibbons, for example, refers to Ireland as "a first-world country, but with a third-world memory" (27). Lloyd draws the same comparison, pointing at Ireland's recalcitrance to modernity and modernization as analogous to that of structurally underdeveloped states in the Third World (384).
9 The concept of combined and uneven development is illustrated in Trotsky's *The History of the Russian Revolution* (University of Michigan Press, 1932). Joseph Cleary, 'Misplaced Ideas?', provides a good discussion of combined and uneven development in the Irish context ('Misplaced Ideas? Locating and Dislocating Ireland in Colonial and Postcolonial Studies' in *Marxism, Modernity, and Postcolonial Studies*. Ed. Crystal Bartolovich and Neil Lazarus. New York: Cambridge UP, 2002).
10 See Seamus Deane's useful Introduction to the *Selected Plays* for the impact of events in Northern Ireland on Friel's drama.
11 On the Field Day Company, see Richtarik.
12 Cleary, *Outrageous Fortune* strongly refutes the claim that modern Irish history is marked by a string of failed political visions. The tragic form of Northern Irish literature has been misread, according to Cleary, as reinstating this vision. I wish to complicate Cleary's argument by using a Vichian materialist framework to uncover the connections between language, form, and the uneven space-time of the Irish periphery in Brian Friel's *Translations*.
13 In drawing parallels between Ireland and the Roman conquest of North Africa, Friel seems to follow the lead of *The History of the Maghrib* by the Moroccan intellectual Abdallah Laroui, who makes a case for extending postcolonial insights into the ancient period.
14 Biodun Jeyifo provides a similar perspective on the question of English in the context of Sub-Saharan Africa, nuancing the earlier positions of the 1970s of which Thiong'o's *Decolonising* serves as a benchmark. I wish to suggest that the resonance between Friel and such African intellectuals as Laroui, Jeyifo, and of course Thiong'o is more than a coincidence; it denotes a dialogue between Irish and other postcolonial currents.

15 See Auerbach's discussion of Vico in *Scenes from the Drama of European Literature*. Lukács, *Goethe and His Age* makes a similarly important connection between the aesthetic achievements of German tragic drama and Germany's relatively underdeveloped condition in the eighteenth century.

Works cited

Auerbach, Erich. *Scenes from the Drama of European Literature*. Minnesota UP, 1984.
Brennan, Timothy. *Borrowed Light: Vico, Hegel, and the Colonies*. Stanford UP, 2014.
—. "The National Longing for Form." *Nation and Narration*, edited by Homi Bhabha, Routledge, 1990, pp. 44–70.
—. *Wars of Position: The Cultural Politics of Left and Right*. Columbia UP, 2006.
Brown, Terence. Ireland: A Social and Cultural History, 1922 to the Present. Cornell UP, 1985.
Cassirer, Ernst. "Descartes, Leibniz, and Vico." *Symbol, Myth, and Culture: Essays and Lectures of Ernst Cassirer 1935–1945*, edited by Donald Phillip Verene, Yale, 1979, pp. 95–107.
Cleary, Joseph. "Misplaced Ideas? Locating and Dislocating Ireland in Colonial and Postcolonial Studies." *Marxism, Modernity, and Postcolonial Studies*, edited by Crystal Bartolovich and Neil Lazarus, Cambridge UP, 2002, pp. 101–124.
—. *Outrageous Fortune: Capital and Culture in Modern Ireland*. Field Day, 2007.
Deane, Seamus. "Introduction." *Selected Plays of Brian Friel*. Faber & Faber, 1984, pp. 11–22.
Dutt, Utpal. *Towards a Revolutionary Theatre*. Seagull, 2009.
Friel, Brian. *Translations*. Faber & Faber, 1981.
Gibbons, Luke. "Ireland and the Colonisation of Theory." *Interventions*, vol. 1, no. 1, 1998, p. 27.
Jeyifo, Biodun. "English Is an African Language—Ka Dupe! [For and Against Ngugi]." *Journal of African Cultural Studies*, August 18, 2017, pp. 1–15.
Laroui, Abdallah. *The History of the Maghrib: An Interpretive Essay*. 1977. Translated by Ralph Manheim. Princeton UP, 2015.
Lloyd, David. "Ireland After History." *A Companion to Postcolonial Studies*, edited by Henry Schwarz and Sangeeta Ray, Blackwell, 2000, pp. 377–395.
Lukács, Georg. *Goethe and His Age*. Translated by Robert Anchor, Howard Fertig, 1978.
O'Toole, Fintan. "Going West: The Country versus the City in Irish Writing." *The Crane Bag*, vol. 9, no. 2, 1985, pp. 111–116.
Pilkington, Lionel. "Irish Theater Historiography and Political Resistance." *Staging Resistance: Essays on Political Theater*, edited by Jeanne Colleran and Jenny S. Spencer, Michigan UP, 1998, pp. 13–27.
Pompa, Leon. *Vico: A Study of the New Science*. 2nd ed., Cambridge UP, 1990.
Richards, Shaun. "Placed Identities for Placeless Times: Brian Friel and Post-Colonial Criticism." *Irish University Review*, vol. 27, no. 1, 1997, pp. 55–68.
Richtarik, Marilynn J. *Acting Between the Lines: The Field Day Theater Company and Irish Cultural Politics 1980–1984*. Oxford UP, 1995.
Thiong'o, Ngũgĩ. *Decolonising the Mind: The Politics of Language in African Literature*. East African Educational Publishers, 1994.
—. *Globalectics: Theory and the Politics of Knowing*. Columbia UP, 2012.
Vico, Giambattista. *The New Science of Giambattista Vico*. Translated by Thomas Goddard Bergin and Max Harold Fisch, Cornell UP, 1984.

9   The heavens look down upon us

José Enrique Rodó and the spirit of América

Marco Katz Montiel

This old man lives simply. He casts off possessions, wears clothing suitable for workers, forsakes a mansion in the centre of the capital for a humble abode on the city's outskirts, drives an outdated automobile, routinely gives away 90 per cent of his income, and devotes his life to improving the human condition. All of these traits would make him remarkable anywhere, but he adds one more to make him truly extraordinary: he governs a twenty-first-century nation.

"All I do is live like the majority of my people, not the minority," the Uruguayan president tells the BBC. "I'm living a normal life and Italian, Spanish leaders should also live as their people do. They shouldn't be aspiring to or copying a rich minority" (Davies). José 'Pepe' Mujica[1] took a more difficult path to executive power than that followed by leaders in Italy, Spain, and most other nations around the globe. The man who would become president spent the 1960s and 1970s with the Tupamaros, an armed opposition group named after a colonial-era resistance figure in Peru and inspired by leaders of the Cuban Revolution. After surviving six gunshot wounds and fourteen years of mistreatment and isolation in prison, he emerged only when Uruguay returned to an electoral system in 1985 (Hernández).

"I'm called 'the poorest president', but I don't feel poor. Poor people are those who only work to try to keep an expensive lifestyle, and always want more and more," he tells the BBC. "This is a matter of freedom. If you don't have many possessions then you don't need to work all your life like a slave to sustain them, and therefore you have more time for yourself" (Hernández). If this sounds at all familiar, it may have the hallmarks of 1960s discourses, which often advocated, as proponents of the Cuban Revolution continue to do, for simpler lifestyles. Even so, practical applications of the complex philosophy expounded by Mujica go back to an earlier period of Uruguayan history.

Twentieth-century postcolonialism, after all, began in precisely this small space. As the first light of a new century beamed down upon an ageing world, a modernist from the margins of America reached back to classical philosophers in the centres of European power to formulate a literary view of decolonization that would reverberate through succeeding decades. José Enrique Rodó transformed his readings of Old World philosophers into concepts suitable for social and political activity in a New World with *Ariel*, an extended essay that explained how young people could learn to undo negative aftereffects of British culture in

America. In this literary act, Rodó turns a 'borrowed light' back on the United States, a hegemonic power that had, in its own struggle for independence, taken that implement of enlightenment on loan from European sources.

"The light that casts its glow on the twentieth century as a whole derived from earlier sources: specifically, from ideas and attitudes I trace back to Vico," writes Timothy Brennan in his explanation of that illuminating phrase. "In time," he continues, "these ideas were taken up by anticolonial thinkers outside Europe and returned to Europe once more, via that influential detour" (*Borrowed Light* 3). Entering into the spirit of Giambattista Vico, supplemented with readings of Georg Wilhelm Friedrich Hegel, Rodó created *Ariel* as a means of speaking back to evolving forms of empire.[2] Although the author employed concepts disseminated by philosophers who famously offered strategies that opposed brutal conquest, his *Ariel* speaks from a different location than those occupied by his predecessors and speaks back to a more surreptitious imperialism than the violent sort experienced by earlier subjects of European domination. From Bartolomé de las Casas to Domingo Faustino Sarmiento, a long history of social and political writings in Spanish had begun with complaints to the Spanish crown and moved to polemics aimed at neighbouring American governments, all the while maintaining the posture of victims fending off violent attacks. Rodó's transformed readings of selected European philosophers specifies American concerns while also seizing control of the discourse among nations struggling to mentally as well as physically release themselves from the colonial condition.

This chapter considers the surprisingly sturdy fruits of Rodó's seemingly fragile transplants in new terrain, first by analysing the peril threatened by Washington's growing empire. After establishing that historical context, I move on to Rodó's remedies: a counterintuitive set of prescriptions, at least to twenty-first-century thinking, that argue for a comprehension of spirit necessary for the expression of the will of the people, an adherence to a state vital to concepts of freedom, and a turn to cosmopolitanism as a means of defence against attempts at hegemonic reification. As a conclusion, this chapter returns to contemporary Uruguay to demonstrate how, in spite of historically misinformed criticism, Rodó's words continue to impact lives throughout America and around the world.

All of these ideas function synergistically in a manner explained by Timothy Brennan. Following Vico's twentieth-century successor, Edward Said, Brennan pulls philosophy, sociology, and politics together in *Borrowed Light*, a philological study that provides "a deliberately generalist understanding of language and literature—a theory of the social that is reliant on a theory of reading based on evidence, correspondence, and evaluation, situating authors in their motives and times" (10). In the context of his time, Rodó's work takes into account Vico's admonishment, "Doctrines must take their beginning from that of the matters of which they treat" (92) or, as Raymond Guess pithily reframes this in the twenty-first century, "political philosophy must be realist" (9). Philologically speaking, Guess argues, "political philosophy should become more historical, or, rather, it should recognize explicitly that it has always had an important historical dimension that, to its cost, it has tried its best to ignore" (69). In line with these philological methodologies, we will begin with the historical moment in which Rodó situates *Ariel*.

Postcolonial America poised for a new form of conquest

By 1900, most American colonies had achieved independence from England, Spain, Portugal, Holland, and France and, following internecine battles, had coalesced or divided into a variety of national units. The danger confronting these new nations in the first decades of the twentieth century, and which would continue to menace nations in Africa and Asia as they gained independence from their own colonial masters later in the century, came not from armed forces, although those remained ready to intervene when all else failed, but instead from cultural practices, which sometimes, due to their seemingly anodyne or even benevolent appearance, had the power to crush the will of peoples in an enduring manner rarely achievable through mere gunplay.

Two years before the publication of *Ariel*, a brief Spanish–American War fixed an Anglo-American image as the dominant force in a hemisphere largely populated by descendants of people who had been colonized by or had colonized on behalf of Iberian monarchies. That dominating part of America came to be known as North America even though this appellation often excludes the United States of Mexico (Estados Unidos Mexicanos)—except, of course, for those states taken by the other United States as part of the 1848 Treaty of Hidalgo—and, most curiously, that remnant of the British Empire, the Dominion of Canada, to this day ruled by the Queen of England.[3] In short, North America referred to a nation that had no real name. America was, and is, a continent, and the United States of America could just as easily serve as the name of any American nation composed of states. Regionally, these particular united states took pride in images of Confederacy, Yankee Ingenuity, and a Wild West Frontier. Oddly, few took any interest in the fact that more of this territory had once belonged to Spain than to England. As Boone points out in *The Spanish Element in Our Nationality*, this historical omission facilitated the fiction of a selectively chosen North America as the true inheritor of the British Empire, a nation ready to take on the white man's burden.

Attempting to unite the white Christian power structures governing the newly independent nations of America, President James Monroe formulated a policy of hemispheric unity in 1823 by declaring,

> as a principle in which the rights and interests of the United States are involved, that the American continents, by the free and independent condition which they have assumed and maintain, are henceforth not to be considered as subjects for future colonization by any European powers.
>
> (Monroe, 'Transcript')

The Monroe Doctrine, Boone reports, "was favorably viewed as anti-imperialist by the new Latin American republics, and even Simón Bolívar, one of the fathers of Latin American independence, imagined a pan-American union that might bring together all the nations of the hemisphere" ('The 1910 Centenary Exhibition' 196). Sadly, later events proved that Monroe had merely overturned one prevailing hierarchy without in any way dismantling the system of hierarchies in America. While calling the political systems practised by the European powers waging war among themselves "essentially different in this respect from that of America," Monroe paradoxically presided over a slaveholding power that, as Sara Johnson points out, collaborated with an axis of Spanish, French, and British armed forces intent on "subduing nonwhite enemy combatants" (67). Other American nations that separated from their European colonizers had the same interest in privileging those regarded as the pure descendants of Europeans. Rather than develop a politics to serve all Americans, observes Benedict Anderson, the initial impetus driving independence movements "in such important cases as Venezuela, Mexico, and Peru, was the *fear* of 'lower-class' political mobilizations: to wit, Indian or Negro or Negro-slave uprisings" (48). Even this limited pan-American unity failed to last; ultimately disdaining collaboration with insufficiently white and Protestant neighbours, the United States (which name we will use from here on to distinguish the most powerful of the various united states of America) affirmed its pretensions to Continental control by effectively spurning José Martí's invitation to participate in the re-formation of America.

An example of these pretensions to Continental control can be seen in the history of relations between the United States and a nation on the opposite end of the hemisphere. In *Para leer el Pato Donald: Comunicación de masa y colonialismo*,[4] an extended essay published in 1972, Ariel Dorfman and Armand Mattelart provide an entertaining—and, as subsequent events demonstrate, horrifyingly prescient—view of the stakes inherent in unequal battles for cultural supremacy between the United States and less powerful nations such as Chile, which has its own history of territorial aggression based on white supremacist notions of manifest destiny (Beckman). Competition between the two began when "Chile had the most powerful navy during the 1880s, and the United States felt threatened"[5] (Boone, '*Una cualidad*'). Although it won wars, most spectacularly conquests of Mapuche territory and the War of the Pacific waged against neighbouring Peru and Bolivia, the power of Chile declined—in part due to a bizarre talking duck. A coup against Chile's elected president in the year following the publication of Dorfman and Mattelart's book received well-publicized support from corporate powers that flouted US law with the connivance of government officials. Outside of Chile, less publicity has been given to the massive public support inside the country for taking down Salvador Allende and subsequently installing the dictatorial Augusto Pinochet, a political environment that continues to hold sway in much of that conservative country. It turned out that the United States did not need troops, tanks, and aircraft to ensure a regime friendlier to its economic aims when a public infatuated with cartoon animals willingly carried much of the burden.

"Children still bear the brunt of marketing campaigns psychologically calculated to influence their elders as they endeavour a secure future for their offspring," observes Mujica in an address to the United Nations General Assembly. Internationally distributed films and comic books and their affiliated merchandise, which includes pillows and plastic figures emblazoned with images of cherished Disney characters, tear down national boundaries, not in some transparent scheme for compatible global coexistence, although it may convey that appearance, but rather as a far more cunning means of assuring the prevalence of a dominant culture in the everyday lives of other nations.

Rodó's remedies: spirit

Eschewing unidimensional polemics, Rodó acknowledges certain benefits derived from the United States. "Any severe judgment made about the Americans of the North must begin by rendering to them, as would be done with all worthy adversaries, the chivalrous formality of a salute," he writes. "My spirit easily complies with this. Remaining unaware of their defects would not seem as foolish as denying their qualities"[6] (*Ariel* 50–51). These qualities, as Rodó sees them, consist of the utilitarian aspects of nineteenth-century English philosophy lamentably unadorned by any spiritual and cultural refinements of the British Isles. Having been the first to cast off European colonialism, the Americans of the north retain the experience of liberty, which they use logically and in practical ways to achieve immediate goals. "And for my part, you see that, although I do not love them, I admire them"[7] (53). This very admiration, he observes, makes it all the easier to imitate their flaws as well as their useful traits.

Admiration for and adoption of traits from one side should not create excuses for wholesale obliteration of another side's culture. Keeping a sharp eye on cultural exchanges, shrewd observers can determine how they sometimes seem deceivingly even; cultural practices have flowed north as well as south, causing an appearance of reciprocal exchange. On inspection, however, currents flowing north are either unacknowledged or, more frequently, accepted in ways that minimize their transformative powers. For example, while the Western Hemisphere Institute for Security Cooperation (WHINSEC) at Fort Benning in Georgia provides military training for Central American soldiers, bureaucrats in Washington, DC reciprocate by taking salsa dance lessons. Weapons and televisions change lives in the south while people in northern climes mostly care about their southern neighbours as cheap labourers to build those devices or as purveyors of simple entertainments that merely add spice to trendy formulations of multicultural experience with smatterings of music, dance, and exotic cuisine.

Responding to the rise of the new colonist on the block, Rodó calls on the Machiavellian protagonist from an allegory of the New World to summon up his religiously faithful spirit, Ariel. "That afternoon," he writes, "the old and venerated master, habitually called Prospero with an allusion to the wise magician of *The Tempest* by Shakespeare, bade farewell to his young disciples congregated around him once more after a year of scholarly pursuits"[8] (*Ariel* 11). Rodó's Prospero offers a clear thesis: in spite of the past and present, the times to come belong to those who develop inner strength.

All who devote themselves to propagating, in contemporary America, a disinterested ideal of the spirit—art, science, morals, religious sincerity, a politics of ideas—should train their will in the persevering cult of times to come. The past belongs entirely to the arms of combat; the present belongs, almost entirely as well, to the rough arm that levels and constructs; the time to come—a future brought closer whenever the will and thinking of those who anxiously await it are energized—will offer, for the development of superior faculties of the soul, stability, location, and milieu.[9]

(68)

Readers can see how the protagonist of *Ariel* deals with the concept of freedom in complex and sometimes elusive ways that do not involve freeing people physically, as many freedom fighters advocate, as much as setting them loose spiritually. The proper state of humans for Vico "consists of mind and spirit, or, if we prefer, of intellect and will" (110). Before dismissing this as unfashionable religious exhortation, readers need to pay attention, as well, to Hegel's insistence on this as part of "the *substantial World* of Spirit" (*Philosophy of History* 33). In *Borrowed Light*, Brennan acknowledges that the employment of terms such as "Spirit" must "sound crypto-transcendent to most readers today" (91) until they understand that this concept does not rely upon gods or any natural laws prepared "to deliver a chiseled tablet of divine commands, thereby establishing our mode of existence" (92). Instead, these philosophical concepts charge us with the responsibility of continually discovering our spirit and then fighting for the realization of that as the will of the people. As with Vico and Hegel, any comprehension of Rodó will depend on an acceptance of intellectual powers capable, however inexplicable this seems, of working *in this world* to go beyond those powers wielded by armed forces and the cruder manifestations of obvious physical reality.

The state

Crucially, for Rodó's purposes, Hegel describes "the shape which the perfect embodiment of Spirit assumes—the State" (*Philosophy of Right* 31). This Romantic adherence to statism as an element crucial to the assertion of a national will that ultimately allows for the greatest cultural expression and individual freedom sometimes makes Hegel difficult for postmodern readers to accept. Responding to the *Philosophy of Right*, Brennan focuses on Hegel's

> unusual and highly original defense of the state form divorced from any particular state (including, of course, the Prussian one). This nexus—of state authority, coercive means, and legal restrictions—more than any other vexes our current thinking about the political, in the humanities no less than the social sciences.
>
> (*Borrowed Light* 84)

Even so, Rodó's employment of this reliance on national identity provided a platform for decolonizing aspirations that remains potent in the twenty-first century. Indeed, in spite of all of the ideals of international amity expressed and attempted since the beginning of the twentieth century, manifestations of culture, including regional offerings, still tend to arise and coalesce within national boundaries, and in some of these new nations, devotion to the state developed as a useful tool to counteract encroachments by larger states. For Hegel, writes Brennan in *Borrowed Light*, "the state is not simply a mechanism for enforcing abstract rights; rather, it is obliged to agree in content with the 'inner necessity' of the good." In this, "Hegel firmly rejects the commonsense assumption that politics is an imposition of external entities onto the natural embodiment of self-expression and freedom" (90). Not understanding human will as immanent in political processes, rather than the reverse, contemporary commentators might view such obeisance as antithetical to individual liberty, but Rodó's reading of Vico and his successors suggests ways in which these governmental entities, when properly ruled, could make possible freedom for all of their citizens.

The most splendid of Rodó's innovations consists in his formulation of American utility in doctrines that did not always function superbly on their native European soil. Writing in Christian climates and thus invested in a particular Holy Spirit, Vico and Hegel chart the development of freedom from the individual through the family and city to the state in a manner that leaves open the potential to embrace all people or, as Vico formulates: self, family, city, nation, all. "The ethical substance, as containing self-consciousness which has being for itself and is united with its concept, is the *actual spirit* of a family and a people," declares Hegel. Tellingly, he adds, when dealing with the actuality of spirit, "individuals are its accidents" (*Elements* 197). Although ethicality, like everything human, begins with individuals, it cannot become, cannot actualize, without societal—or, as one might dare update the language of this concept, socially conscious—unity. In *Phenomenology of Spirit*, Hegel develops this thought by establishing culture as a set of communal practices that endow people with a spirit capable of bringing people together in systems while maintaining their individual nature (298).

This concept shines through in *Ariel*, in which Prospero gathers his disciples, forming them into a group meant to wield combined power even while each remains substantially distinct. Such formulations will not likely receive a fair hearing from listeners, including those professing a rejection of capitalism, indoctrinated in freedom based solely on the exercise of consumer choice. The free will of people depends first on the "immediate or *natural* ethical spirit—the *family*" and then on the establishment of good governance or, as Hegel puts it in a phrase that demands a reading in at least two senses, "in the *constitution of the state*" (*Elements* 198). And the importance of this constitution, however read, can be seen in the fact that it need not result in good governance; even bad governance serves better than none. Offering a warning heeded by Rodó, Vico inveighs against the absence of laws, as in all-powerful monarchies or "unlimited popular commonwealths," because these lead to the greatest tyranny, "for in them there are as many tyrants as there are bold and dissolute men in the cities" (87). Comprehended in this fashion, individual freedom simultaneously leads to and arises out of national power; whatever its constitution, a state must be constituted.

None of this reliance on governance, good or bad, should make citizens complacent, however; the state constitutes not a final goal but instead a platform from which to begin fusing the elements that will lead to new truths and freedoms. Vico urges a continuous questioning that Hegel will transform into a dialectic art. In keeping with his providential understanding of a Holy Spirit, the Neapolitan philosopher discusses "the term 'divinity' [i.e., the power of divining], from *divinari*, to divine, which is to understand what is hidden *from* men—the future—or what is hidden *in* them—their consciousness" (102). Rodó brings this to life by having Prospero regale his disciples with the tale of an ancient king who hospitably opened his immense palace to the people (*el pueblo*). Even travellers from other lands were welcomed to pass through the monarch's unguarded entrances and take in the splendorous chambers. But one part of the king's fortress remained hidden from view; visitors did not even know of its existence and no one but the king could legally step inside. Once ensconced in this secret chamber, the king, now transformed into an egotistically ascetic figure, heard no sound and smelled no odour from the outside world. While the outside world continued to regard him as open and generous, the king would periodically disappear into his redoubt to be carried away by "the white wings of Psyche"[10] (25), the mythical princess associated with the human spirit. In this secret space, Death finally approaches to remind the king that he has only been one more guest in his own palace. His remains never leave the inner chamber "because nobody would have dared set an irreverent foot in that place where the old king wanted to remain alone with his dreams and isolated in the ultimate Thule[11] of his soul"[12] (25).

"I relate this story to the scene of your interior realm," concludes Prospero:

> Open, with a healthy liberality, like the house of the trusted monarch, to all of the world's currents there exists at the same time the mysterious hidden cell, unknown to run-of-the-mill guests, where nothing other than serene reason belongs. Only when you penetrate this inviolable vault will you be able to really call yourselves free men.[13]
>
> (25)

This inner space, set up by the author as something to be sought within oneself as well as in the places maintained by those in power, dangles in front of Prospero's listeners and Rodó's readers. The protagonist of *Ariel* does not offer to reveal this secret location, but urges his disciples to discover it. In Book IV of the *New Science*, Vico discusses "The Course the Nations Run," concluding with a section on governmental reliance on secrecy and mystery. Moving from dealing with "extreme savagery," when "religion was the only means sufficiently powerful to tame it," Vico takes his readers through the progress of governance to "popular commonwealths, which are naturally open, generous, and magnanimous" and thus "went on to make public what had been secret" (350). In *Ariel*, Rodó expresses the hope that his students will form part of the changes that move from dictatorial theocracy to credible democracy.

A call to cosmopolitanize

This discussion of the ancient king's interior realm reminds us of the need for knowledge in the fight against hegemony. Although the subjugation of America included horrific deployments of armed force, the continuing oppression of the Continent relied on weapons of mass culture. From the missions of California, venerated by tourists and other faithful followers, to California's most famous duck, sermons, books, periodicals, and screens have induced people to ignore their spirit and allow their states to prostrate themselves before a foreign master. In part, Rodó finds the solution in an embrace of cosmopolitanism. As Brennan points out, Vico understands the benefits found in "the rise of the settled, class-ridden world of agricultural estates on the grounds that individual realization is not possible outside of the cities' protections, however ugly these first forms of government were" (*Borrowed Light* 26). Although Vico attaches a caveat to the increase of personal property in relatively few hands, he realizes, as Brennan observes, how "the ethical leap that leads to the establishment of laws relies on the creation of settled communities" (77). At a time when many North Americans still longed for Romantic notions of country life, Rodó clearly saw how these could prove harmful when directed at Americans living in the south.

To this day, in fact, a parochial stance that purportedly lauds people by focusing on supposedly natural characteristics easily slips into notions of folkloric tribes incapable of governing themselves in the modern world. An insistent focus on what seems 'natural' continues to reinforce endogamous tendencies by perpetuating myths of a cultural consciousness exclusively available to communities considered originary. Certainly Latin America, however defined, has more to offer than 'world music' and trendy restaurants. As an example, Brennan's *At Home in the World* points to Alejo Carpentier's acclamation of popular culture "exported globally by a country (unlike the United States) that had no imperial claims. The importance of this view," he adds, "is inseparable from the almost universal assumptions that exist about the supremacy of U.S. mass culture as a political foundation for a new globalism" (298). Widespread dissemination of Cuba's popular culture leads to questions that Brennan poses concerning

the possibility of a viable mass culture in noncapitalist countries—countries, in other words, where the subaltern is nominally in power. It raises the question, for example, of whether a corporate, technological mass culture has displaced the authentic, collective forms gestured to in the testimonial, and it makes one wonder whether a type of exportable mass culture—that is one capable of global dissemination—exists differently from the varieties now made in the United States and (to a lesser extent) Western Europe.

(303)

The answers to Brennan's questions do not always encourage hope. In addition to the aggressive marketing of culture allied with a lack of serious attempts at cultural education, audiences in the United States have typically made little effort to move outside of their comfort zone, leaving spectators with superficial impressions of Latin America that may cause more harm than would complete ignorance.

"How 'cosmopolitan' has been used negatively in the past adds to the built-in stigma of so using it today," writes Brennan, who recalls its derogatory "meanings as earlier applied to 'Christians, aristocrats, merchants, Jews, homosexuals, and intellectuals'" (*At Home in the World* 20). Unwitting imperialists maintain images of the south as impractical, emotional, and unprepared for world affairs or even self-governance. Randolph Pope observes that scholars almost feel an obligation to denigrate Latin American governance and infrastructure. "Too much celebration risked being naïve, reactionary, and even a form of betrayal of the writers' responsibility to their continent," he believes, as he laments how a gloomy outlook encourages "the general tendency, especially in the US, to concentrate teaching and research on the Latin American dismal: dictators, poverty, dirty wars, corruption, and so on" (137). Acknowledging a number of sound reasons for Latin American gloom, Pope makes what should seem an obvious point regarding the way "dreadful schools and unemployed workers do not magically stop north of the Río Grande" and how anyone visiting the major cities of Chile, Argentina, Mexico, and other Latin American nations cannot help noticing how "they easily dwarf most cities in the US not only in development, but also in opportunity and sophistication" (138). And this is not new. As Rodó pointed out in a reference to London and other great cities of the world, "There are already, in our America that is Latin, cities whose material greatness and apparent aggregate of civilization approaches them at an accelerated pace to participate as first-class in the world"[14] (67). These places exist even if smug imperialists cling to ignore-ance.

Rodó wants all Americans to embrace this cosmopolitanism, which, in his eyes, does not exclude feelings for the past or links to ethnicity or tradition, but instead endures as part of a cultural inheritance "of elements that will constitute the definitive American of the future"[15] (50). Propounding this view, Rodó joins the many philosophers who have, over millennia, considered great cities "a necessary organism of high culture" and "the highest manifestations of the spirit"[16] (65). Recent scientific studies add weight to Rodó's assessment. "Compared with metropolitan areas, a higher percentage of deaths occurring in nonmetropolitan areas from the five leading causes among those aged <80 years were potentially excess deaths," concludes a 2017 analysis by the Centers for Disease Control and Prevention (Moy et al.). What members of this research team call "excess deaths" are those attributed to rural life, a potential killer when compared with urban existences in developed countries. In a powerful demonstration that rural life affects more than a person's physical condition, recent electoral results in the United States clearly reveal elevated levels of inchoate anger and paranoia linked to deflated levels of formation among populations that vehemently reject cosmopolitanism.

Ariel in the twenty-first century

Long recognized as an important anti-hegemonic tract of its era, *Ariel* conveys far more useful information than generally credited it for outlining future possibilities as well as the perils of powerful mores. Without resorting to the stances of either beleaguered sufferers or criminal vengeance seekers, positions adopted, as we shall presently see, by scholars such as Roberto Fernández Retamar, Rodó offers nuanced observations on interactions between unequal powers and demonstrates how old philosophies can be put to new uses in contests for national identity. Although smaller countries resisting hegemony have often proved spectacularly unsuccessful, the methods outlined in *Ariel* have worked to the advantage of some Davids fighting their own Goliaths and, as seen in one of the first Uruguayan presidents of the twenty-first century, continue to offer a means of resistance. As we have seen in this chapter, in order to comprehend—and, more urgently, utilize—Rodó's text as an enduring vital literary document, contemporary readers need to know the historical context in which it was written, understand the multidirectional complexities of hegemony and cultural appropriations preceding and succeeding the publication of *Ariel*, and grasp the movement of its philological bases between Europe and America.

Spirit forms the core of Rodó's message, a spirit formed by the will of the people that leads to freedom for nations. Will, not wilfulness, makes up this expression of free spirit; rather than the independence demanded by angry heirs of complacent classes, the patterns of individuality practised by fashionable bohemians, or the shattering of all institutional regulation by disillusioned souls impatient with what they perceive as intractable elitist systems, this freedom finds its foundation through a social cohesion that later in the twentieth century enables the successes of Tupamaro guerrillas (among them a future president of Uruguay), the revolutionary religious adherence of converts such as El Hajj Malik El-Shabazz and Muhammed Ali, the Young Lords of East Harlem, and the current rise of Black Lives activists in the very heart of the empire.

Responding to his literary predecessor, Fernández Retamar abuses the spirit who assists Prospero in *The Tempest* with a term occasionally employed in public discourse as an epithet: intellectual. Artfully situating the despised word in a Gramscian setting, Fernández Retamar brutally bifurcates Ariel's options:

> he can, as have the anti-American intellectuals, decide between serving Prospero—with whom he is apparently on the best of terms, but with whom he will never go beyond being a fearful servant—or join with Caliban in his fight for real liberty.[17]

(72)

As Brennan points out, "resistance to Hegel in postwar anticolonial circles takes many forms." In an example that could apply to a postwar author such as Fernández Retamar, some argue "that the master/slave dialectic is a metaphor of class struggle without any particular reference to actual slavery and therefore a diminishment of the specificity of racial oppression" (*Borrowed Light* 99). Rodó does not lose sight of that specificity, making his text's subjection to Fernández Retamar's bifurcation untenable. Instead, the author of *Ariel* offers remedies for a particular situation at the beginning of the twentieth century.

In this assessment of Rodó's views, it is fair to acknowledge that Fernández Retamar finds support from Shakespeare's script, in which Prospero treats the benevolent Ariel with nearly as much abuse as he inflicts on the malevolent Caliban. Ariel suffers because he does not follow orders unquestioningly: "Remember I have done thee worthy service, / told thee no lies, made thee no mistakes, served / Without or grudge or grumblings. Thou did promise / To bate me a full year" (22). This reasonable assessment earns nothing better than the protagonist's protracted scorn, with Prospero recalling how he saved Ariel and then going on to call his servant names and threaten to return him to his previous state of torture. Thus, in Fernández Retamar's retelling, Caliban must attain his rightful place in the conduct of revolution, particularly when the situation has become entirely hopeless and raping the master's daughter followed by other acts of death and destruction offer the only hope for change, even if the new order will do no more than switch the roles of the players in a possibly worse regime. This type of violent overthrow generally appeals to men who have never tasted the blood of battle. Like Plutarch and other Hellenic philosophers who bequeathed Europe their aggressive posture, they remain in thrall to myths of a communal Spartan purity that gives the impression of right made by might. Efforts to make right with might have left history with such a horrifying legacy of human suffering that leaders should only consider such bellicose engagements as a last resort, a recourse that, even in the most extreme cases, might be discarded when annihilation could prove preferable to culpability in reciprocal atrocities. No. Before invoking Caliban, Ariel must be given his—or her—due.

Mujica did not need to wait as long as Rodó for his anti-intellectual comeuppance. Writing in *The New Republic*, Eve Fairbanks brings up the disappointment of Uruguay's left, and calls the former president's dreams "too good to be true." Presented as "a president who was not only a living antidote to the culture of materialism—'You don't stop being a common man just because you are president,' he told the *Guardian*—but also an extraordinarily eloquent advocate for those same principles, a philosopher king without modern parallel," Mujica, as Fairbanks sees it, ultimately failed.

> When I started to read about Mujica, I noticed that few of the many articles written about him considered the tangible effects of his tenure on his country. Was he able to create the deep change he calls for in his speeches? Last year, I went down to Uruguay to find out. I booked two weeks in the country and scheduled more than two dozen interviews. But it only took a day or two before the Mujica myth began to come apart.

Needing only a day or two out of her hardly ambitious plan for two weeks of research, Fairbanks appears to have little trouble unearthing Uruguayans who admire Mujica as a person but dislike him as a president. These assessments charge the nation's leader with building expectations based on a style that never realizes the implementation of its substance. Perhaps most seriously, Fairbanks and the citizens with whom she converses fault Mujica for the public's failure to adopt their president's attitudes about materialistic lifestyles. If anything, they claim, conspicuous consumption has risen rapidly since Mujica's inauguration.

Such assessments of the president of Uruguay's time in office fail to take into account the long-term effects of a leader committed to social evolution. In his speech at the United Nations, Mujica hearkens back to Rodó and the beginning of the twentieth century, when Uruguay "placed itself in the vanguard of the social, the State, education. One could say that social democracy was invented in Uruguay"[18] (Mujica). He looks forward as well and urges the assembled nations to consider an abolition of borders through methods not based on intemperate ducks, Donald or otherwise. Still, he admits to feelings of anguish over "the time to come that I will not see, and for which I have committed myself. Yes, a world with better humanity is possible, but perhaps for now the first task should be taking care of life"[19] (Mujica). Uruguay is one of a handful of countries—others include India and South Africa—that have produced leaders ahead of their times. The words—and the actions based on those words—of these few globally recognized heroes of the twentieth century continue to influence people inside and outside of their countries. As even the most casual glance at daily headlines shows, useful change develops more slowly than impulsive gestures made to please the basest notions of a mob, but it does develop. Like Rodó, Mujica speaks to generations to come, with a message to young people taken from Prospero's discourse to his followers in *Ariel*. The spirit of the populace requires time to develop.

With sincere hope for something the *Oxford English Dictionary* defines as "unlikely to be fulfilled" (keeping in mind that *unlikely* does not mean *impossible*), *Ariel*, like Vico's *Scienza nuova*, concludes with a call for piety (Vico 426). Following Prospero's speech in *Ariel*, the disciples gather outside. The youngest looks up and speaks:

> As the crowd passes, I see that, although they do not look at the sky, the heavens look at them. Over their dark and indifferent mass, like furrowed ground, something descends from above. The vibration of the stars looks like the movement of a sower's hands.[20]

> (Rodó 74)

Vico begins his book with an "Explanation of the Picture Placed as Frontispiece to Serve as an Introduction to the Work." In this close reading, the ray that descends from heaven and then reflects "from the breast of the metaphysic onto the statue of Homer" (5) forms the author's first demonstration of the borrowed light that will shine through centuries later in Brennan's work. The light remains light even as refraction alters it, just as the philology of Vico continues to illuminate when Rodó changes its direction. The power of these ideas resides in their ability to change completely even as they remain the same.

Notes

1 Pepe, a common nickname for José, also refers to common people.
2 Among the numerous references to philosophical thought in works by José Enrique Rodó, references to Vico appear in *Motivos de Proteo* and to Hegelian thinking in *Ariel*.
3 Although some casual observers insist that this monarchical hangover has nothing more than symbolic meaning, which in itself should trouble democrats, the Queen of England, who also reigns as the Queen of Canada, retains the power to prorogue Canada's parliament. As Duff Conacher reports, the Queen's representative has already invoked this mighty power on the federal level three times in the twenty-first century, once to close debate on a bribery scandal (2003), again to prevent a duly elected prime minister from assuming office (2008), and finally to cover up the treatment of Canadian armed forces detainees in Afghanistan (2009).
4 Later translated as *How to Read Donald Duck: Imperialist Ideology in the Disney Comic*.
5 "Chile tenía la marina de guerra más poderosa durante la década de 1880 y Estados Unidos se sintió amenazado" (this and all other translations my own unless otherwise indicated).
6 "Todo juicio severo que se formule de los americanos del Norte debe empezar por rendirles, como se haría con altos adversarios, la formalidad caballeresca de un saludo. Siento fácil mi espíritu para cumplirla. Desconocer sus defectos no me parecería tan insensato como negar sus cualidades."
7 "Y por mi parte, ya veis que, aunque no les amo, les admiro."
8 "Aquella tarde, el viejo y venerado maestro, a quien solían llamar Próspero, por alusión al sabio mago de *La Tempestad* shakesperiana, se despedía de sus jóvenes discípulos, pasado un año de tareas, congregándolos una vez más a su alrededor."
9 "Todo el que se consagra a propagar y defender, en la América contemporánea, un ideal desinteresado del espíritu -arte, ciencia, moral, sinceridad religiosa, política de ideas-, debe educar su voluntad en el culto perseverante del porvenir. El pasado perteneció todo entero al brazo que combate; el presente pertenece, casi por completo también, al tosco brazo que nivela y construye; el porvenir -un porvenir tanto más cercano cuanto más enérgicos sean la voluntad y el pensamiento de los que le ansían- ofrecerá, para el desenvolvimiento de superiores facultades del alma, la estabilidad, el escenario y el ambiente."
10 "las blancas alas de Psiquis."
11 In Greek mythology, the northernmost part of the world.
12 ". . . porque nadie hubiera osado poner la planta irreverente allí donde el viejo rey quiso estar solo con sus sueños y aislado en la última Thule de su alma."
13 "Yo doy al cuento el escenario de vuestro reino interior. Abierto con una saludable liberalidad, como la casa del monarca confiado, a todas las corrientes del mundo, exista en él, al mismo tiempo, la celda escondida y misteriosa que desconozcan los huéspedes profanos y que a nadie más que a la razón serena pertenezca. Sólo cuando penetréis dentro del inviolable seguro podréis llamaros, en realidad, hombres libres."
14 "Existen ya, en nuestra América latina, ciudades cuya grandeza material y cuya suma de civilización aparente las acercan con acelerado paso a participar del primer rango en el mundo." Note that Rodó properly employs the lower-case L when using Latin as an adjective rather than a proper noun. For this reason, I render this portion of his text as "our America that is Latin" rather than "our Latin America" in my translation.
15 ". . . de los elementos que constituirán el americano definitivo del futuro."
16 ". . . un organismo necesario de la alta cultura . . . las más altas manifestaciones del espíritu."
17 ". . . puede optar entre servir a Próspero—es el caso de los intelectuales de la anti América—, con el que aparentemente se entiende de maravillas, pero de quien no pasa de ser un temeroso sirviente, o unirse a Calibán en su lucha por la verdadera libertad."

18 "... se puso a ser vanguardia en lo social, en el Estado, en la enseñanza. Diría que la socialdemocracia se inventó en el Uruguay."
19 "... el porvenir que no veré, y por el que me comprometo. Sí, es posible un mundo con una humanidad mejor, pero tal vez hoy la primera tarea sea cuidar la vida."
20 "Mientras la muchedumbre pasa, yo observo que, aunque ella no mira al cielo, el cielo la mira. Sobre su masa indiferente y oscura, como tierra del surco, algo desciende de lo alto. La vibración de las estrellas se parece al movimiento de unas manos de sembrador."

Works cited

Anderson, Benedict. *Imagined Communities: Reflection on the Origin and Spread of Nationalism*. Revised ed. and extended ed., Verso, 1991.

Beckman, Ericka. "Imperial Impersonations: Chilean Racism and the War of the Pacific." *E-misférica*. vol. 5, no. 2 Web, Winter 2008, http://hemisphericinstitute.org/hemi/en/e-misferica-52/beckman.

Boone, M. Elizabeth. "The 1910 Centenary Exhibition in Argentina, Chile, and Uruguay: Manufacturing Fine Art and Cultural Diplomacy in South America." *Expanding Nationalisms at World Fairs: Identity, Diversity and Exchange, 1851–1915*, edited by David Raizman and Ethan Robey, Routledge, 2017, pp. 195–213.

—. *The Spanish Element in Our Nationality*. Pennsylvania State UP, 2019.

—. "*Una cualidad lírica de un encanto duradero*: La pintura norteamericana y chilena en el Centenario de Chile en 1910." Museo Nacional de Bellas Artes, 2014.

Brennan, Timothy. *At Home in the World: Cosmopolitanism Now*. Harvard UP, 1997.

—. *Borrowed Light: Vico, Hegel, and the Colonies*. Stanford UP, 2014.

Chiquita Brands L.L.C. "The Chiquita Banana Jingle." 2016. www.chiquita.com.

Conacher, Duff. "Proroguing Parliament without Cause? Canadians Want It Banned." *The Globe and Mail*, August 23, 2013, www.theglobeandmail.com.

Davies, Wyre. "Uruguay Bids Farewell to Jose Mujica Its Pauper President." March 1, 2015, www.bbc.com/news.

Dorfman, Ariel and Armand Mattelart. *Para leer el Pato Donald: Comunicación de masa y colonialismo*. Siglo XXI Editores, 2001.

Fairbanks, Eve. "Jose Mujica Was Every Liberal's Dream President. He Was Too Good to Be True." *New Republic*, February 5, 2015, http://newrepublic.com.

Fernández Retamar, Roberto. "Calibán." *Calibán—Contra la leyenda negra*, prologue by Carmen Alemany, Ediciones de la Universitat de Lleida, 1996.

Guess, Raymond. *Philosophy and Real Politics*. Princeton UP, 2008.

Hegel, Georg Wilhelm Friedrich. *Elements of the Philosophy of Right*. Translated by H.B. Nisbet, edited by Allen W. Wood, Cambridge UP, 1991.

—. *Phenomenology of Spirit*. Translated by A.V. Miller, Oxford UP, 1977.

—. *The Philosophy of History*. Translated by J. Sibree, Batoche Books, 2001.

Hernández, Vladimir. "José Mujica: The World's 'Poorest' President." *Magazine*, BBC Mundo, Montevideo, November 15, 2012, www.bbc.com/news/magazine.

Johnson, Sara E. "'You Should Give Them Blacks to Eat:' Waging Inter-American Wars of Torture and Terror." *American Quarterly*, vol. 61, no. 1, March 2009, pp. 65–92.

Martí, José. "Nuestra América." *El Partido Liberal*, Mexico City, March 5, 1892. Accessed December 13, 2009.

Monroe, James. "Transcript of Monroe Doctrine (1823)." National Archives and Records Administration. www.ourdocuments.gov. Accessed February 11, 2017.

Moy, Ernest et al. "Leading Causes of Death in Nonmetropolitan and Metropolitan Areas—United States, 1999–2014." *Morbidity and Mortality Weekly Report (MMWR)*, vol. 66, no. 1, Centers for Disease Control and Prevention, U.S. Department of Health and Human Services, January 13, 2017.

Mujica, José. "Discurso completo de Mujica en la ONU." *Diario La República*, September 27, 2013, www.republica.com.uy.

Pope, Randolph D. "Antonio Skármeta's Uniqueness." *A Contracorriente: A Journal on Social History and Literature in Latin America*, vol. 10, no. 1, Fall 2012, pp. 124–146.

Rodó, José Enrique. *Ariel.* 1900. Linkgua Ediciones, 2008.

—. *Motivos de Proteo.* Valencia: Editorial Cervantes, 1918.

Shakespeare, William. *The Tempest.* 1611. Bedford, 2009.

Vico, Giambattista. *The New Science of Giambattista Vico.* Translated by Thomas Goddard Bergin and Max Harold Fisch, Cornell UP, 1984.

10 Historicizing language and temporality in José María Arguedas' *Deep Rivers*

Mela Jones Heestand

José Arguedas' 1957 novel, *Deep Rivers*, depicts its fourteen-year-old protagonist Ernesto in his travels through the Peruvian Sierra followed by his internment and eventual escape from a Catholic boys' school in the southern central town of Abancay. As Ernesto navigates complex and often contradictory racial, class, and gender codes of the 1920s Sierra region, he remains stalwart in his solidarity with the oppressed indigenous groups he encounters. From the omniscient narration, we learn that Ernesto's affinity for indigenous culture and his wish to belong more fully to the indigenous community relate to his treasured (and sometimes invasive) memories from early childhood when he received the loving care of indigenous servants and then spent time living with a Quechua '*ayllu*'.[1]

However, rather than indulging in an unproductive and melancholic nostalgia, as some critics of the novel have contended, I will argue that in certain processes of memory, the author's articulation of the 'then' in the 'now' points towards the kind of postcolonial possibilities for a transformative future that Timothy Brennan invokes in *Borrowed Light* (2014). Brennan's recuperation of Vico's non-presentist form of historicism, which he argues prefigures Marx, will anchor and fortify my analysis of the novel's materialist treatment of language and memory, which manifests in Quechua linguistic and aesthetic practices suggesting the potential to reinvigorate the present moment through continued contact with and integration of past experiences. This, in turn, points to vital modes of imagining individual growth and collective change as dialectically interrelated.

Although Arguedas' novel grounds temporality in the social forms unique to Peruvian colonial and postcolonial experience, *Deep Rivers* nonetheless remains relevant and historically connected to 'Western' strains of Marxist thought. In fact, I would argue that it is precisely through the unfolding of a Vichian philological approach that we can best understand the unique possibilities as well as the limitations implied by Arguedas' affective, sensual, and intersubjective treatment of language and temporality. Based on the logic of "intellectual generalism" and interdisciplinarity developed in Left Hegelian thought, as Brennan describes it, the Vichian approach allows us to draw on multiple disciplines, without privileging one over the other (5). In this mode of intellectual litheness, we can begin to peel back and examine the mediations that structure the subject/objects of inquiry in *Deep Rivers*.

From the perspective of linguistics, Arguedas puts on full display the dynamic possibilities implied by certain kinds of Quechua verbal and other forms of aesthetic practices. On the other hand, as Marxist linguist V.N. Voloshinov (1929) explains it, the word, and indeed any ideological sign, is Janus-faced. It may have the capacity "to register all the transitory, delicate, momentary phases of social change, because the process of its materialization is dialectical" (19). However, the "social multi-accentuality of the ideological sign" through which it remains vital, dynamic, and capable of further development is also related to the social dynamic that makes it "a refracting and distorting medium" (23). Because the sign does not merely reflect but also refracts—as it continually passes through individual psyches to its materialization as speech and back again—this means that it intersects with "differently oriented social class interests within one and the same sign community" (23). For this reason, Voloshinov writes that the ruling classes "strive to impart an eternal character to the ideological sign" in order to "extinguish or drive inward the struggle between social value judgement which occurs in it, to make the sign uni-accentual" (23). In the case of *Deep Rivers*, the 'Voloshinovian' view of linguistic change must also take into account anthropological explorations of the Quechua worldview, in which multiple dimensions of existence (human, animal, and spiritual) participate in such dialectical transformations of individual and collective consciousness.

Further, as we pursue a 'Quechua-ized' Voloshinovian perspective of historical change regarding how the 'then' enlivens the 'now'—pointing *not* towards pathological regression on an individual or collective level *but rather* towards the possibilities for making and re-making ourselves—I will also explore the historically determined ways in which these processes become inhibited, or perhaps even impossible, in Arguedas' novel. Drawing on Freud's exploration of the unconscious, I will pursue a materialist analysis of the individual and social boundaries and their spatial logics that function to freeze the dynamic mechanisms of change. I will argue that rather than holding universalized ideas of how language structures human consciousness, or of how an individual integrates into his/her social milieu, scholarship must pursue with tenacity and flexibility, as Vico and Marx did, the historically determined relationships among language, consciousness, and the social forms in which they are embedded.

Representing Quechua language as stasis

The early critics of *Deep Rivers* found fault with the novel's perceived technical clumsiness, but it was later recuperated in the 1970s and 80s from the dustbin of the 'primitive regional novel' and considered alongside other innovative Latin American 'boom' novels. Even so, many critics who inspired renewed appreciation for *Deep Rivers* nevertheless considered its temporally backwards orientation to be an unproductive one. For example, Latin American literary critic, Antonio Cornejo Polar, sees protagonist Ernesto's inward journey as a return to the paradise of his prior immersion in the life of the Quechua *ayllu*, when and where he maintained "a conception of the universe, understood as a coherent and compact totality that was absolutely integrated" (my translation) (100). Even though Cornejo generally valorizes indigenous culture and the power of the protagonist's memory to rescue him from the threat of psychological disintegration, Ernesto's resistance to letting go of the past "is his most grave limitation" (108).

Estelle Tarica's much more recent intervention into the novel is similar, but her focus shifts the terms of Cornejo's critique, focusing on the power dynamics involved in Arguedas' representation of indigenous culture. Rather than presume the historical stagnation of indigenous culture per se, which Cornejo does to some extent, for Tarica, the immobile domain of indigeneity lies within the author's consciousness. She reads *Deep Rivers* as a kind of anti-*Bildung*, in which the narrator/protagonist successfully resists his coming-of-age by clinging to what she claims coincides with Arguedas' feeling of possessing an "inner" indigeneity. Ironically, he is able to assert such a definition of indigeneity because of his position at the top of the racial hierarchy as an upper-middle-class mestizo intellectual.

Tarica's analysis also takes aim at the recuperation of *Deep Rivers* by certain critics, such as prominent Latin American literary critic Angel Rama, her sharpest criticism revolving around what she views as hegemonic representations of *mestizaje*, or racial/cultural mixing. Rama contends that Arguedas acts as novelist, but also ethnographer, anthropologist, and folklorist, mediating 'from the inside out' the indigenous world for the reader, and he thereby achieves the ultimate *gesta* (feat) of the mestizo. Tarica warns us that this type of reading risks reproducing, rather than challenging, racist norms, and she cautions against maintaining a blind optimism that a highly empathic, boundary-crossing mestizo could gain "intimate knowledge without power and that ethnography could operate without its colonizing function" (82).

Rather, she points to evidence suggesting Arguedas' withdrawal from contemporary indigenous cultural life and, in the writing of *Deep Rivers*, his flight into a "disembodied" linguistic netherworld of Quechua language as a symbolic system. In this way, the writing of the novel represents the author's own inner conflict as a mestizo intellectual who makes questionable claims as to his relevance and belonging to indigenous culture. Key to her argument, Tarica draws our attention to Rama's paradoxical position that, as an individual with insider status in the indigenous community, Arguedas was capable of mediating between "regional interior" and "external-universal" and of articulating the mythical thought of a Quechua people holding 'magical' beliefs about language. At the same time, she reminds us, according to Rama, Arguedas also held Saussurean views regarding the strict and unqualified disjunction between the signifier and the signified.

Tarica points out that one cannot simultaneously believe that language has special powers of connotation and denotation *and* that signifier and signified have no logical internal relationship. She suggests that Arguedas performed a sort of mental compartmentalization, wherein he viewed the Spanish language within a modern Saussurean context, while Quechua language "consisted for [him] of particular kinds of verbal experiences, most especially onomatopoeia, that he believed could occur only in Quechua as a consequence of its particular linguistic nature" (88). And if, for Arguedas, Quechua words have intrinsic denotative and connotative meanings, Quechua language is therefore closely related to Quechua song, which can even more effectively convey meaning through the intensified modulations of the voice.

With this in mind, Tarica analyses a pivotal passage in the novel in which the omniscient narrator identifies himself directly with his young protagonist. Here, the narrator writes, speaking of a bird while the protagonist stands waiting in the streets of Abancay for some girls to arrive,

> Its song transmitted the secret of the deep valleys . . . *Tuya, tuya*! While I listened to its song, which was surely the material I am made of, the diffuse region from where I was torn and thrown amongst men, we saw the two girls appear in the orchard.
>
> (my translation; 348)

Tarica interprets this region of birdsong as the "unbounded, diffuse and immaterial region of sound" that represents Arguedas' "understanding of Quechua culture . . . where/when there is presence yet not identity" (95). Furthermore, she contends that the kind of "translation as explanation" of Quechua sound systems that Arguedas performs throughout the novel functions to create a temporally stagnant and de-materialized version of Quechua culture that the author uses to describe his own inner state as an alienated but spiritually "limpid" mestizo intellectual (348). She argues that the interconnected wholeness of the domains of Quechua culture is contrasted to the bounded domains of "temporal earth," the world of men where corrupting knowledge and original sin threaten the protagonist's spiritual limpidity, a term, she tells us, that Arguedas himself used to describe the Andean soul. This disembodied and static version of Quechua culture *as* language (or rather, as a disembodied linguistic system) represents a place of refuge from the (embodied) social integration and maturation that one usually finds in a coming-of-age novel; and, Tarica argues, it ultimately has little to do with the experiences of indigenous people struggling to assert their sense of cultural identity in a complex and diglossic society. In contrast, according to Tarica, Arguedas is interested in the purity of an indigenous culture that lies within, which he stamps with authenticity because of the childhood experiences he claims to have had.

As relevant as this intervention is in regard to the novel's neglect of the complexities of indigenous Andeans navigating the difficulties of a diglossic society, Tarica's assessment of the novel's temporal stagnation leaves untouched certain assumptions about the supposed universality of what constitutes "healthy" social and psychological boundaries and integration. Moreover, if we accept the author's assumption of the polarization of psychological and social realms in her positing of a strictly differentiated 'inner' and 'outer' language/culture, we are left with the question of exactly how these realms are mediated and what this means in terms of how cultures do, in fact, intermingle. In other words, even if *Deep Rivers* depicts the contours of an inner 'indigenous' self, this internal world must still communicate itself to the outward, socially oriented mestizo who narrates the novel to the reader. And if it is simply the case that the 'outward' aspect of the mestizo expresses what he believes to be his true inner essence through the writing down of "translation as explanation," as Tarica suggests, this presents two problems.

First, if the socially outward mestizo assumes control and domination over the content of his inner "magical" self, these contents cannot but lose all traces of their power of enchantment. Second, Tarica's critique overlooks key portions in the novel in which the inner and outer worlds of the narrator/protagonist become severely dislocated, and time does indeed appear to stagnate. Ironically, I will argue that these moments of pathological disjunction of 'inner' and 'outer' are precisely what provides us with the kind of depiction of the psyche and of subjectivity that Tarica implies in her analysis, where an 'inner self' is rigidly controlled and bounded by an 'outer self'. More broadly, the indigenous 'inner' mestizo intellectual is channelled, managed, and sublimated through the linguistic practice of novel writing.

Examined more closely, this depiction, in turn, corresponds to specific processes of memory and development that follow late-Freudian models, wherein the psyche is bounded and the individual's sense of autonomy is shielded by linguistic means. Language, in this instance, is meant to safeguard the individual both from excessive external excitation as well as from the inner contents of the unconscious; in this way, it constructs the boundaries consonant with 'healthy' individuation. Within these implicit views of the psyche and its relationship to language, we will see that it is difficult, if not impossible, to conceptualize a way for the novel's young mestizo males to 'come of age' without emotionally distancing themselves from the brutal treatment of marginalized 'others', which they are being groomed to perpetuate. Along these same lines, if we accept the universality of the late-Freudian psyche and its relationship to language, it is seemingly impossible to theorize about any sort of cultural mixing that does not reproduce hegemonic power imbalances from the point of view of the marginalized culture in *Deep Rivers*.

El patio interior: spatiality 'beyond the pleasure principle'

Oddly ignored by the majority of the novel's critics, perhaps the most disturbing scenes in *Deep Rivers* take place in the boys' school's inner courtyard (*el patio interior*), located beyond the main courtyard (*el patio de honor*) and between the kitchen and the adobe wall separating it from the playing fields. At the far end of the inner courtyard, a wooden fence conceals a row of makeshift toilets. It is in this space of the inner courtyard, under the cover of darkness, where a mentally challenged woman kept by one of the priests makes her appearance, and where the older boys drag her into the toilet stalls and sexually assault her, while the younger boys watch. The actions and reactions of the boys are recounted in episodic detail over the course of many scenes scattered throughout the novel, accumulating in our sense of the inner courtyard as a claustrophobic space whose temporality is defined by a repetition compulsion.

Despite the terror that the younger boys experience when witnessing the older boys' sexual assaults on '*la demente*', the narrator/protagonist says, "But I, too, many afternoons, went to the inner patio behind the big boys, and I contaminated myself watching them" (94). Ernesto describes the smells "that oppressed us" along with the quality of the light at the time of evening when '*la demente*' made her appearances: "The walls, the ground, the doors, our clothes, the sky at this hour, so strange, without depth, like a hard roof of golden light; everything appeared contaminated, lost or wrathful" (95). The boys and the world itself seem to emanate helpless despair, "like river fish when they fall into the turbid water of mudslides" (91).

However, while Ernesto's feelings of dissociation from his surroundings and his elevated levels of fear and arousal may strike the reader as an understandable response to the horrifying events, the alienating force of this cycle of abuse multiplies if we consider that these scenes are, in part, a re-telling by an older and presumably wiser narrator looking back on a childhood trauma that nevertheless seems to have retained its full emotional charge. For the protagonist and his peers, it makes sense that they would not become sensitized to the abuse and would remain on high alert in the inner courtyard, even if the threat of repetition of the violence were only slight. From the point of view of the narrator, who at times provides extremely lucid ethnographic and historical analyses in the novel, the lack of narrative distancing from the protagonist's pain and bewilderment only adds to the confusing terror of this spectacle. Where the abuse of '*la demente*' is concerned, the reader gets no sense of why the boys desire this particular kind of sexual experience, why the abuse is permitted by the priests, and why the boys feel they cannot stop. It is as if the older version of Ernesto—at times so obviously separate from the young boy who participates in the novel's action—becomes engulfed in the horror and cannot exist in the moment of narration. The narration itself feels caught in an out-of-control cycle of dissociation and hyper-arousal, and the reader has no way to fully comprehend the protagonist/narrator's 'inner' world, because the 'outer-oriented' narrator is completely incapable of explaining it.

In the next section I will look at the structural relationship between the inner and outer courtyards, as well as the implied relationship between 'inner' turmoil and 'outer' acting out in terms of trauma. I will argue that in the case of the traumatized individuals pictured in these scenes, there is no clear division between the psychological and the social. In fact, the peculiar linguistic and temporal stagnation that emerges appears to reveal an uncanny isomorphy between the structuration of inner and outer realms, which in turn points towards their historically determined, rather than universal, nature. The temporal breakdown in these scenes goes hand in hand with the narrator's odd inarticulateness, whose memories seem so oddly fresh with intensity, while at the same time we get a sense of why the protagonist cannot 'come of age', since it would seem to require him to adapt to the cruel and oppressive social norms. These are precisely the psychological and social norms that torment him in the first place.

The Freudian psyche

Regarding what I see as the novel's peculiar psychological and social isomorphy, Freud's final representation of the psyche presents us with certain striking structural features if we confer on the psyche a horizontal, rather than vertical—as it is typically viewed—orientation. Examined horizontally, the visual depiction of the relationships between the conscious and unconscious realms of the psyche Freud presents, which we will examine shortly, bears a remarkable structural similarity to the layout of the boys' school. In *Beyond the Pleasure Principle*, published in 1920—shortly before the time in which the events of the novel would have taken place—Freud attempts to account for the puzzling phenomena that he refers to as the "repetition complex." Based on his observation of children's play as well as his therapeutic reflections on patients who had suffered trauma or felt themselves under the daemonic spell of "fate," Freud asks why these individuals compulsively repeated unpleasurable experiences. This requires Freud to re-think his previous topographical model of the psyche, which he had conceptualized based on the idea that individuals act in accordance with the pleasure principle in so far as it does not conflict with the reality principle. Taking into consideration what he saw as the flouting of the pleasure principle, Freud refines his idea of how the psyche functions.

Freud describes as a "system of energy" his final and best-known structure of id, ego, and super-ego, where the ego rests above the id and adjacent to the super-ego, which resides in the upper-left tier of the psyche. Positioned as it is, between the unconscious realms of the id and super-ego, the ego mediates the flows of energy in their surges of excitation and repose. But whether the ultimate goal of the psyche is "to free the mental apparatus entirely from excitation" or to form excitatory object-relations remains unresolved in Freud (625). Rather, we are left with a dynamic model of constant push and pull in which the sexual drive of Eros motivates the organism to form object-relations with others and, in contrast and also in response, a death drive directed at the self and then turned outwards comes from the wish to return to an inanimate state, or a zero-level of excitation. Once at rest, the self is meant to reanimate and seek out object-relations again, and so on.

The Eros/Thanatos dialectic, in turn, raises the question for Freud of whether feelings of pleasure and unpleasure can be produced equally from what he refers to as "bound and unbound excitatory processes" (626). Unbound excitatory processes are those related to the so-called primary drives associated with the infantile need for the mother, drives that split off during the formation of ego and find their home in the aggressions of the id. According to Freud, "there seems to be no doubt whatever that the unbound or primary processes give rise to far more intense feelings," leading to his claim that the pleasure principle must have been present before these high levels of excitation began to bombard the individual as an infant. The struggle of the child is one of binding or cathecting these primary excitatory processes, which involves "attaching psychic energy . . . to an object, whether this is the representation of a person, body part, or psychic element."

The developing child begins to build his/her ego boundaries through word-presentations that comprise the 'thing'—a primarily visual representation of integral gratification—plus the presentation of the word associated with it. Word-presentations work in networks of other word-presentations, "connecting paths between ideas, without being led astray by the intensities of those ideas" (340). It is the child's acquisition of oral language, therefore, that ultimately helps form the protective structural barriers to the overwhelming intensity of the drives by subordinating them to processes of the mind responsible for "attention, judgment, reasoning and controlled action" (340).

Turning our attention back to the school, we can note that, laid out flat, Freud's ego bears a structural similarity to the outer courtyard where the boys enjoy themselves and do not 'act out' violently towards each other or on '*la demente*'. The outer courtyard abuts the external world of the school and to one side are positioned the rooms of the priests, isomorphic with the super-ego. If the school grounds functioned properly—as a structure devoted to the healthy development and maturation of the boys—the space of the outer courtyard and the supervision of the priests should be able to keep their sexually violent behaviour in check. The narrator describes the paved outer courtyard as a place of contentment and light, where the boys with a proclivity for performance play spirited mestizo songs and tell stories.

But even for the boys for whom music and storytelling have a special draw, like Ernesto's friend, Romero, and Ernesto himself, the main courtyard cannot hold their interest powerfully enough to stave off the terror of their urge to enter the inner courtyard:

> In the paved [outer] courtyard, where we sang funny, gay *huaynos* [traditional Peruvian songs], where we talked peacefully and listened to and told interminable tales of bears, mice, pumas and condors . . . never could we be free from sudden attacks of fear of that [inner] courtyard.
>
> (58–59)

If we think of the outer courtyard in terms of its function of binding and sublimation through language, in particular the intensified language of song, the space is not powerful enough to mediate and hold back the dark forces that lie further within. As we shall see, what lies within the inner courtyard—and within the shadowy inner-depths of boys who lust for violence—has nothing to do with 'inner' indigeneity.

On one hand, if we follow Freudian theory, the dictates of the reality principle here are fuzzy enough to allow for the abuse of '*la demente*' to continue. Freud argues that the social standards of behaviour of a given culture determine the dictates of the super-ego, associated with the unconscious dread of the father's punishment. In that sense, the boys need not have had a guilty conscience, since the priests are strangely absent during the loud and tumultuous episodes in which the older boys yell and run into the toilet stalls all together, jockeying for the chance to assault her. As acting authority figures, the priests could and should have put a stop to the violence, and as the adult men with whom the boys would have had a super-egoic identification, they should have provided a positive example for the boys to internalize. However, not only does it seem that the priests look the other way while the abuse is happening, there is even reason to believe they are also sexually exploiting '*la demente*', as the boys can observe her going in and out of one of the priest's rooms.

Whereas the boys manage not to dwell much on the violent acts they commit against one another, the sexualized nature of '*la demente*'s' abuse brings them face to face with the deeply painful humiliation of feeling devoid of self-governance, even while they are perpetrators of egregious acts of objectification. With bodies that feel completely separate from the direction of their 'souls'—or in Freudian terms, instinctive drives that are not adequately reined in by ego and super-ego—they behave as condemned body-objects, as though they are under daemonic possession. The boys appear to Ernesto "like goblins, like monsters who appear in nightmares, moving their hairy arms and legs" (95). The profound psychological disruption that results from these incidents has its foundation and corollary in the breakdown of the social fabric of the school, and leads us to a more specific historical discussion of why the boys might be perpetrating these acts in the first place and why they compulsively repeat them. Neither symbolic language nor the literal presence of disciplinary father figures can cathect, bind, and/or stop the behaviour that traumatizes everybody involved.

Against universalizing notions of language and psyche in postcolonial thought

As we discussed earlier, Tarica reads the protagonist's inability to join his social group of developing mestizo young men as related to the narrator's continued problem of clinging to his 'inner' indigeneity. Neither the adolescent protagonist nor the adult narrator can join his social group in a healthy, responsible, and mature way, because both are lost in 'inner' worlds. In the case of the protagonist, Ernesto remains stuck in an 'inner' world of nostalgic memory, whereas the narrator/author has constructed a false indigeneity, based not on lived experience but rather on an un-enlivened notion of language/culture as a set of abstract, grammatical principles. In this line of thinking, language, as it is portrayed in *Deep Rivers*, does not play an active, integrative, and mediating role between individual and the social form.

This sort of strict polarization of the domains of 'inner' and 'outer' as they relate to language implies a Spivakian logic, especially if we consider Tarica's assessment of Arguedas' first novel, *Yawar Fiesta*. Here, she argues that Arguedas' depiction of the inarticulateness of the bilingual experience more suitably represents the "lived indigenous experience" in its discursive conflictedness and enigmatic nature. *Yawar Fiesta* portrays a sort of misfire, and thus a more apt representation of the way in which one might speak 'for' the subaltern, in the Spivakian sense (111). As Neil Larsen explains, in this all-too-familiar circularity of a certain kind of postcolonial framework—whose foundational critic was Homi Bhabha—cultural hybridity is theorized as a site through which other denied knowledges "enter upon the dominant discourse and estrange the basis of authority" (114). As Larsen puts it,

> "the colonial presence" it would appear, is to be "resisted," "subverted," "estranged," and even, perhaps even "changed" entirely through a practice of exposing it for what it really is and always was . . . Colonialism exposed is colonialism overcome; its doing becomes its undoing.
>
> (38)

Within this framework and its implied relationship between 'inner' and 'outer' discursive domains, indigeneity *should* be located *not* inside the consciousness of the author/narrator or a novel's characters, but rather in a realm of discursive exteriority and inscrutability.

In contrast, we have seen the suggestion that the world inhabited by Ernesto uncovers the historically determined nature of the late-Freudian psyche, precisely because it is neither fully internalized nor consciously externalized. Rather, the structures associated with the regulation of drives fail internally and externally in the same kinds of ways, leading us to speculate that the social form *and* the psyche must be subject to historical transformations. Furthermore, the possibilities regarding 'outer' changes of the social form and the 'inner' changes of the psyche might well be tightly interrelated.

Similarly, I will argue in the next section that—just as we miss the dialectic possibility of individual and collective change if we divorce psychological and social dimensions in the novel—we must also grasp the dynamic field of relationality between 'inner' and 'outer' language. In other words, if we consider linguistic structure as supra-individual and separate from the social, as strict Saussurean models of language do, we fail to see the thought and social forms in which Quechua language and culture are embedded. Quechua language, precisely in its capacity to change structurally, functions at a level that is both keenly individual and profoundly social. In the next section we will look at the ways in which certain kinds of linguistic and aesthetic experiences ripple through individuals and collectivities.

The *zumbayllu*: a symbol of harmony or fragmentation?

As I suggested earlier, the Vichian tradition lends support to our historicization of language as a mediator between 'inner' and 'outer' realms, as we attempt to bypass the dead-end application of 'universal' categories to 'postcolonial' works that articulate alternative forms of temporal change, as well as accounts of temporal stagnation. Vico's forebears, the Neo-Platonists, belong to this thread, as they strove to "speak of submerged Afro-Asiatic inheritances, hiding in the light borrowed from Egyptian and Phoenician civilization" (Brennan 11). Just as they confronted the orthodoxy of established Christianity that set to "rupture lines of connection between thought on the peninsular continent of Europe and Egyptian and Levantine learning," Vico rejected Cartesian dualism, which "pitted human essence against social existence" and "challenged a newly confident Enlightenment rationality in order to understand foreign cultures on their own terms" (Brennan 42).

Hegel, Brennan tells us, also belongs in this theoretical tradition, having accomplished perhaps the greatest disruption to Eurocentric fixity by making thought itself an object of intellectual inquiry. Hegel, Brennan explains, does not insist on a prescriptive kind of thinking "but rather seeks to capture in words how thought proceeds as such." For Hegel, knowing results from self-questioning, "wherein one watches one's own thinking as it gets around earlier misconceptions, placing itself in contact with an external ground." Truth comes about in a "sensuous encounter whereby the Notion adequates itself by painful stages of opposition to the inner necessity of otherness" (109). Through a process of "negating our earlier, merely inward selves in a dynamic encounter with existent things, we are not simply applying concepts to object from outside them, but self-reflexively, coming to understand their reality by witnessing the work of our own" (109). It is in this way that we can understand history as the "inventory of human labour congealed in cultural artefact as the material warrant of Mind" (76).

In the case of *Deep Rivers*, the lines of thought that run between the Quechua worldview of linguistic and identitarian fluidity and those strands of the Vichian tradition that reached early twentieth-century Peru—via George Sorel's influence on José Carlos Mariátegui—certainly impacted Arguedas. Arguedas was guided in his thought by Mariátegui's unorthodox approach, wherein he believed the power of myth could awaken the revolutionary potential of the people. Inspired by Mariátegui, Arguedas participated in the creation of a renewed Peruvian literature "in which indigenista advocacy would pave the way for indígena agency" (García 18). In the case of *Deep Rivers*, however, the most vital "borrowed light" emanates from Quechua thought, providing us with a viable alternative to the dysfunctional universalized understandings of the psyche of capitalist 'post'-colonial modernity. Unlike social and thought forms that seem unable to free themselves from the anxiety of Eros, the 'magic' of indigenous/mestizo socio-cultural forms gives us a glimpse at powerful and erotic modes of creating relationships of complementary difference. These relations in turn imply alternative ways of understanding the role of language and aesthetics in movements of collective change and individual maturation. In the Freudian terms in which we spoke in previous sections, we can also begin to see that so-called "bound" and "unbound" processes associated with need and desire do not have to become alienated. Neither is it universally true that needs and desires are competitive and individual matters. Rather, the desired and desirable needs of the community can also be regulated collectively.

The *zumbayllu* as mythic/ritual object

As alluded to earlier, much of the previous analysis of *Deep Rivers* focuses on the extreme complexity of the cultural mixing that occurs in this novel with a strong emphasis on how the Spanish and Quechua languages come together. To recap Polar's interpretation of the *zumbayllu* as a symbol of cultural mixing, Arguedas' paragraphs introducing this enchanted object give a rapturous series of linguistic associations of Quechua words that incorporate the onomatopoeic morpheme, *-yllu*; these include a buzzing insect, Quechua instruments, and a scissor dancer whose syncretic and "diabolical performance" at a Saints festival unite elements of Christian and indigenous religion. The Spanish '*zumbar*', also onomatopoeic for buzz, clearly constitutes the principal root of the word. And this toy is not only syncretic by way of linguistic interrelationships. For one, Ernesto and his friend Antero believe that the young mestiza girl, Salvinia, possesses all of the beautiful qualities of the *zumbayllu*. The *zumbayllu* also appears to the have the capacity to bridge literal and social distances, much as music does in the novel.

On the other hand, mixed characters and hybrid symbols also speak to the extreme social fragmentation of Andean society. Antero shares Ernesto's belief in the sentient nature of the river, and thus he also shares aspects of an indigenous view of the cosmos in which human beings are intimately related to the natural world. However, Antero states quite blatantly that although as a child he felt compassion for the Indians when they were being whipped, he is more than willing to do the same when he becomes a landowner. And as Cornejo Polar puts it, the *zumbayllu* comes to symbolize "a painful reality: a broken universe" after Ernesto's rift with Antero. Ernesto learns that "brotherhood and hatred are simultaneous" and "universal brotherhood impossible" (*Los Universos Narrativos* 127). Since Antero has become an enemy of the Indians and an abuser of the young girl he fell in love with earlier in the novel, Ernesto can no longer be his friend. Cornejo Polar writes that this rupture destroys the *zumbayllu*'s magic.

In essence, the novel's constant reinsertion of violence and hatred undermines the valorization of cultural 'hybridity' as a linguistic model of mixing morphemes together to come up with new combinations. As Tarica so clearly points out, it leaves unanswered the question of who exactly is subsuming whom and into what sort of psycho-social matrix. However, from the point of view of Voloshinov, Cornejo Polar addresses the reflective capacities of language, but not its refractive ones. Nor does this mode of linguistic analysis allow us to contemplate fully the possibilities for social transformation that the novel engages, because it fails to take into account Quechua social and thought forms that would help us to understand more specifically the kinds of inter-relation that it suggests.

In the chapter entitled 'The *Zumbayllu*', just after an extensive scene devoted to the performance of Andean music that Ernesto witnesses during his surreptitious wanderings into the merchant-class (indigenous) mestizo section of Abancay, the protagonist poses a question of fundamental importance: "what are people for?" (181). Before leaving the bar, he tells an indigenous musician whom he recognizes from his travels, Don Jesús, to sing "before the cross. For me, so that I may leave here soon" (181). A coastal regiment has just descended on Abancay in response to an (indigenous) mestiza-led rebellion, and when Ernesto returns to the city centre to meet the army band marching towards the plaza, his attention turns to some crickets, killed en masse by people "with no consideration for their sweet voices, for their inoffensive, graceful figures. They killed a messenger, a visitor from the enchanted surface of the earth, when they could have let it fly away" (182).

Finally, Ernesto notices "the strollers" and ponders the strangeness of people from the coast and their difference from Andeans. These events, which on the surface may appear unrelated, all bring Ernesto's question—"what are people for?"—into sharp focus. What are people for, or what effects can their words and actions have in the world; can the indigenous musician's singing really precipitate Ernesto's departure from Abancay? Certainly, he seems to be resisting what often appears to him to be the case: that humans are on earth for the senseless purpose of causing suffering to non-humans (the crickets, for example), and to one another. Even while there is much to suggest the contrary, Ernesto also rejects the idea that people exist "for" the purpose of suffering, and that suffering is the basis of our common humanity. When he listens to the school rector's sermon to the *colonos* (the indigenous peasants indentured to the town's landowners), for example, Ernesto clearly apprehends that the rector's glorification of their suffering not only reinforces the violence of the status quo, but it also blots out of consciousness more powerful forms of relationality.

As Cornejo Polar sharply points out, Christianity conceives of sin as something that de-links man from God, but Ernesto sees sin as separating him from the sociality of nature and the world, such that on those nights after '*la demente*'s' abuse, he felt as though he had fallen into a crevasse, "where no voice or encouragement from the busy world could reach me" (62). This view entails a distinct conceptualization of the relationship between body and soul, culture and nature and, in considering these relationships, we can begin to unfold the corollary questions implied in Ernesto's wondering to himself, "what are people for?"—namely, what alternative models of personal and social transformation does *Deep Rivers* entertain?

As Amerindian anthropologist Viveiros de Castro (1988) explains it, difference, as seen through the logic of 'Western' thought, has to do with a metaphysical discontinuity. A primitive disunity is believed to distinguish humans from animals as well as 'inferior' humans, despite a physical continuity that links all body-objects, all of which are ruled by the necessary laws of biology and physics. In other words, we are all alike, despite our mental and spiritual differences, because we suffer similar bodily experiences. In contrast, Viveiros de Castro argues that in Amerindian thought, non-humans differ from humans not because they lack culture or because their soul is non-existent or qualitatively distinct, but rather because their bodies differ. This bodily difference is not so much physiological, but one of "affects, dispositions and capacities" that render a species unique and influence the way beings see different things in the same way (478).

Thus, the nature/culture split that separates animals from humans in Amerindian thought is not ontological and defining; it is a matter of a being's point of view. Moreover, the primary relationship between humans and animals is social, requiring humans to manage the mixture of humanity and animality in animals and to differentiate the many natures out of a universal sociality. Viveiros de Castro writes that "the manifest form of each species is an envelope concealing a human form" only visible to shamans and other beings with unique perceptive abilities (471). The anthropologist Michael Uzendoski stretches the definition of Amerindian shamanism to include certain types of aesthetic activity by individuals who are not actually shamans. In the novel, the boys' ecstatic play with the *zumbayllu* appears to be one such ritual "shamanistic" activity.

If difference is rooted in the body in Amerindian thought—in its affects, dispositions, and capacities—and universality in the soul, this does not mean that the body is static, or even that it is distinguished from the soul. (A being's affects, dispositions, and capacities are part and parcel of its point of view, and this point of view, this consciousness, *is* the soul.) Rather, Amerindian thought emphasizes the body's ability to transform and assume the affects, dispositions, and capacities of other beings. Shamanic activity opens the way for bodily transformation of this type, thus enabling the acquisition of another point of view and the establishment of social relations with other subjects. As Aparecida Vilaça explains, "the world is now seen the same way as the new companions, that is, the member of the other species" (351).

For example, singing attracts the power of what is mimicked, such as a bird, and the bird is then felt in the body. The singer then exists not just as individual subject but also as a mimeticized Other: "She/he creates the bridge between the original and copy that brings a new force, the third force of magical power, to intervene in the human world" (Taussig 106). This new force is related to the power of the subjectivities that the shamanic activities have attracted, and thus constitutes a form of subjectivity that does not simply facilitate the manipulation of the material environment. Its centre is not an egoic reality principle, but rather it is able to attend vigilantly to the specificity of any phenomenal entity at hand. And here, phenomena do not become mere exemplars of an imposed classificatory framework—like the pre-conscious system of oral language that quickly organizes all of the developing child's interactions with externality—but rather they are integrated into an ever-expanding inter-subjectivity.

Uzendoski's ethnographic subjects use the powers they attract from birds through their songs to communicate with faraway loved ones and wayward spouses. In much the same way, the *zumbayllu* in *Deep Rivers* functions as an aesthetic object with the ability to attract a number of different subjectivities through mimeticized aesthetic performance. When Ernesto spins the *zumbayllu*, a significant part of this activity is to say the word over and over in a clearly mimetic linguistic performance that syncs up with "the whirring of the top," which "was like a chorus of *tankayllus* [whirring insects]." The narration reads, "It made one happy to repeat this word, so much like the name of the sweet insects that disappear humming upward into the light" (68). The spinning of the top creates image and especially sound, forming a linguistic bridge between those subjects who mimic it and its own mimetic activity, bringing a magical power into play that can affect the human world. And the "bridge" that is created between the original and the copy, the merging of self/other, subject/object calls into being a force that has a profound effect on Ernesto and the other boys. Añuco, one of the school's most violent boys, watching the spinning *zumbayllu* from the edge of the group, "looked like a new, recently converted angel" (67).

As Cornejo Polar points out, the boys gather round this object, and for a brief period of time their conflicts are resolved and a renewed brotherhood informs their relationships. For Ernesto, the *zumbayllu* "was a new kind of being, an apparition in a hostile world, a tie that bound [him] to the courtyard [he] hated, to that vale of sorrows" (69). The power that is unleashed by the *zumbayllu* also alleviates Ernesto's status as an unassimilated stranger in a foreign town—a status that is tantamount to psychological torture in Andean lore—and allows him to forge a connection with the town of Abancay, even if he remains a boy from another region. He is still an outsider, still different, but the spinning of the *zumbayllu* has enabled a productive and complementary relationship with the new town. For Ernesto, the *zumbayllu* also has the ability to erase the social distance that he feels between himself and a young mestiza señorita—a distance that in the past had seemed unbridgeable.

A second *zumbayllu* that one of the boys makes, called the '*wink'u zumbayllu*', has even more remarkable powers than the first. For one, it seems to be able to cross geographical spaces vaster than the first; its voice travels across mountains, rivers, and valleys to reach Ernesto's father in a distant part of the nation. Ernesto even believes the *wink'u zumbayllu* can reach the leader of the (indigenous) mestiza-led rebellion that has shaken the town, as well as the *colonos* at the hacienda and the *ayllu* members (called the *comuneros*) that took him in as a young child. If this were the case, the *wink'u zumbayllu* would be able to cut through and heal a set of race/class animosities that the Spanish colony and the Peruvian State had worked hard to foster—those between indigenous peasants (*colonos*) and partially independent and landholding *comuneros*. In Peru at the time of the novel's writing, the capacity to communicate with and bring together diverse types of human groups—*colonos*, *comuneros*, mestizos, and *blancos*—was potentially even more radical than communication between human and animal subjects.

The '*wink'u zumbayllu*' never fully loses its ability to mediate intersubjective relationships in the novel, even if it loses some of its efficacy. When one of the novices asks Ernesto's friend, Antero, how he has fashioned a toy "to make it change its voice like that," since on one spin this *zumbayllu* merges with the sun and makes the *pisonay* sing (117), Antero answers that it has nothing to do with his actions but rather "the material it's made of." We might say that its affects, dispositions, and capacities influence what kinds of mimetic performances and bodily transformations it can make, and in turn what sorts of powers it can let loose into the world. In this sense, the extreme permutability of the '*wink'u zumbayllu*'s' body means that it can also attract forces that diminish its powers, such as the blessing by a novice, Brother Miguel, along with its contamination in the hands of Ernesto's friend, Antero, after he has forged a friendship with an abusive coastal boy named Gerardo. By the same token, on the morning of '*la demente*'s' death near the novel's end, Ernesto considers digging up the *wink'u zumbayllu* that he has buried after worrying about its potential for ill effects. He thinks to himself: "it might have learned some other sounds, since it has been sleeping underground" (208). In other words, its continued permutability, like that of all bodies, means its contamination might have been reversed by its new environment, making the *zumbayllu* capable once more of attracting the kinds of forces and unleashing powers that could have a positive effect.

The *wink'u zumbayllu* seems almost to exceed the status of aesthetic object, because the boys come to believe that it has a soul, along with much stronger powers of attraction. We might even suggest that it approaches the status of a fetish object, since at times it appears to act as an independent entity, which functions without mediation by human activity and consciousness. Unlike the protest dance that electrifies the town at the beginning of the novel, the '*wink'u zumbayllu*' unleashes an indigenous aesthetics without clear social actors, set into motion in the secretive performances of children who are unable to understand the full implications of what they are doing. Uprooted from the social forms that give meaning to this sort of aesthetic practice, the *wink'u zumbayllu* is only partially controlled by boys who are eventually encouraged to give up such 'childish' games when they note the intensity of Ernesto's 'play'.

Rather than willingly abandon his 'play', Ernesto remains unwilling to 'grow up' and assume the violent and contradictory roles associated with (upper-)middle-class mestizo masculinity. That Ernesto can resist and have an awareness of the processes that influence the *zumbayllu*'s dialectical potential to change and be changed is a testament to the deep-rootedness of indigenous thought and social forms in the Peruvian Sierra at the time. Moreover, it is strongly reminiscent of the still contested "myth of *Inkarrí*" that Flores Galindo popularized in his ethnographic, historical analysis of what he saw as the dynamic of apocalyptic desire that swept across Peru at various points in time.

According to Flores Galindo, for those indigenous communities who believe the myth to be true, the final Inca's head is growing underground, and when it regenerates and becomes reconnected with his body, a new era will begin. During this new era, the world will invert—the rich will be poor and the indigenous people will rule over the 'whites'; those from below will replace those from above. Flores Galindo explains that this myth does not actually stem from pre-Columbian times but rather attests to the merging of European ideas of utopia, certain Christian millenarian impulses, and the Andean notion of *pachakutik*, or world reversal. Nevertheless, this myth has exerted a formidable influence on indigenous social uprising and the indigenous and mestizo imagination throughout the twentieth century and beyond (67–68).

What kind of psyche, what kind of collective?

The idea that the Oedipal schema is anachronistic and needs to be re-examined is not new. As Dagmar Herzog's recent book on Freud points out, both Lacan and Horkheimer asserted that "the father-relationship so central to Freudian theory" was outdated. In *Black Skin, White Masks*, Frantz Fanon also challenged the universal applicability of Freudian psychology for people of colour, discussing the social effects of white supremacy on the developing psychological structures of black children.

Clearly, if we accept my argument regarding the isomorphy between psychic and social spaces in *Deep Rivers*, something is awry. Not only does the social super-egoic corollary fail to fulfil its purpose—from their rooms, the fathers do not watch over the boys and prevent them from harming themselves or others—it also appears that in the psychological process of the binding of this particular trauma, its repetition compulsion, the death wish has no end. In both instances, the key relation has been lost—the relation to father—since it is supposedly the boy's eventual identification with the father that allows him to sublimate anti-social impulses and his metaphoric castration that makes the maturing boy suitable for full citizenship in group life. Ernesto's world is one in which fathers deserve little or no identification, since not only do fathers often fail to love, in the case of the school's father figures, or to protect; in the case of Ernesto's biological father who abandoned him to the school, the fathers have failed to sublimate their own anti-social impulses (sexual and otherwise).

However, the structuring role of the Oedipal schema does seem to retain some of its descriptive power, despite its serious limitations and anachronistic feel. Why, we might ask, in a Vichian philological mode, does this particular 'then' exist in the 'now'? For our present purposes, literary theorist Juliet Flower MacCannell helps illuminate the interrelated social and psychological breakdown we see in *Deep Rivers* in her explanations of why Freud himself "placed the problematic of the modern symbolic under the sign of Oedipus" (28). She writes, "we must conclude, especially after reading *The Ego and the Id* that [Freud] found it dysfunctional under conditions of modernity" (19). MacCannell argues that Freud's later text deals with a crucial displacement: "the unconscious, the id takes over from the father-superego as the model for the collective" (9). Indeed, in the novel, even while their abuse of the '*la demente*' causes the boys to express feelings of deep alienation, paradoxically, the space of the inner patio also appears to be a markedly 'collective' space, wherein the boys behave as a cohesive group in the acting out of their sadistic desires.

Sharing some similarities to the arguments of Fanon, though arguably focused more on the oppressors rather than on the oppressed, MacCannell argues that in the place of what still seems to be the Oedipal psyche, modernity is historically characterized by its shift to "the narcissistic psyche." She cites as reasons for the fall of the Oedipal psyche the toppling of monarchism (God and king), the inauguration of 'liberté, egalité, fraternité' of the Enlightenment, and the widespread dispersion of the economic system of capitalism. She writes that twentieth-century fascism drove the nail in the coffin of our ability to place any faith in the parental form "in whom we can invest libido in perfect confidence" (19). Drawing on the insights of Fanon and others, such as W.E.B. Du Bois, we must also emphasize the catastrophic psychological effects experienced by the victims of the systematic dehumanization that began with colonialism and slavery.

Western society could have done away with the Oedipal form, MacCannell argues, but instead we have dragged it along with us into modernity devoid of its substance, which is "to moderate the 'ego-centred' passions, to civilize and foster and to support sexuality through difference." Under the "name" of the father, MacCannell tells us, "another and sadistic Other—unconscious, superego, It[2]— has begun its reign of pleasure and of terror" (12). This sadistic Other—unlike the overly harsh but still loving and sustaining Oedipal father—looks much more like the 'phallic father' of Freud's *Totem and Taboo*, who was killed and eaten by his sons for his tyrannical excesses as the one and only "possessor of goods and libidinal claims." In essence, MacCannell argues that Freud invents this particular myth to explain humankind's transition from 'primitive life' to 'civilization' in such a way that it reads very much like the 'primal scene' of capitalist modernity—built upon the backs and predicated on the spilled blood of enslaved Africans and indigenous peoples.

The son who replaces the despotic father does not bear his guilt internally "as a debt toward his progenitor" as Oedipus does, because his father was a tyrant. By eating his father, the son internalizes his powers, but, although he enjoys them, "they undergo a displacement via a 'guilt' that is projected outward in the form of denying himself relation to his family members, especially of the opposite sex" (16). Unlike the Oedipal myth in which the maturing boy child becomes unable to relate to his mother whom he has desired sexually— no parental 'aura' of difference holds sway over the brother—after the killing of the phallic father he can no longer relate to his sister: "They [the sisters] are subject to a taboo which takes the positive form of the rule of exogamy," because their sexual difference is the very basis of his identity, his ability to associate himself to the symbol of the phallus (16). MacCannell argues that what looks like modern patriarchy is, in fact, what she refers to as the "Regime of the Brothers," a fraternity that brutally excludes the sister's difference from consciousness, which would lay bare her equal rights in the power grab after the father's death. I would add that it is not only the sister's difference that is excluded, but also that of the racial 'other'.

By "occupying the [white] father's (empty) place, playing his role, the Brother can simulate a 'symbolic order' of law and morality" and imagine himself to be the neuter and neutral umpire. However, this supposedly 'objective' order of judgement is none other than the brother's repressed taboo of sexual—and I would add, racial—difference, and the Brother's symbolic order nothing but a mirroring fraternal 'imaginary' whose (unconscious) 'categorical' imperative is that he enjoy him(self). According to MacCannell, what the Brother ultimately desires is not a sexual (or any other kind of) relation of difference to an(other)—which would require him to attend to the difference of the sister, to what she might desire and the power that is due to her. The Brother wants repose in his own self-same image and to gain release from the attendant conflicts of sexuality and difference.

If, as I explain above, the sons of the 'phallic father' assume the tyrant's powers by killing and eating him, they can only do so debt-free if they project the guilt outward by denying themselves relations with the 'otherness', since this is the repressed relation which authorizes them to identify with the now empty symbol of the phallus. But even though the sons deny their dependence upon the material and spiritual difference of the 'sister'—they separate desire from need and love—this is the very basis of their fraternité. The unconscious desire that connects them together is the desire to destroy the 'otherness' of the female and the racial other, so they can be free to enjoy their self-same image. What MacCannell refers to as the "mad coupling and uncoupling" of modern/post-modern fiction illustrates the pseudo-relation whose aim it is to enact and re-enact the double alienation of need and desire for the other. Paradoxically, because in the novel '*la demente*' represents such a pure physicality—the boys have free rein to do exactly what they please to a body that is deemed not to have a soul (at least in the first half of the novel)—it is impossible to alienate their desire from a need for her, 'couple' as madly as they will. She does not provide the semblance of a relation—the pseudo-sexual relation that MacCannell argues is the "only relation that seems to exist now [in capitalist modernity]"—because her very marginalization and vulnerability makes her difference too extreme to repress. Whereas the señoritas with whom Ernesto and the other schoolboys are playing at courting have the power to refuse or even to 'desire' the subjugation of their difference—they have been presented with the 'choice' of the upper-class mestiza woman of whether to identify with the male psyche or to 'elect' the absence of identity—'*la demente*' cannot choose anything. And because they cannot repress their collective drive to commit acts of abject, sadistic violence, they do not 'enjoy' this experience of 'fraternal mirroring', and nor do they perceive it as social. On the contrary, they experience an excruciatingly 'true' reflection of their 'inner' selves as anti-social monsters.

From the point of view of the novel's narrator, there does not seem to be a way to master this material and translate it from a primary realm of terrifying drives to a secondary domain of the control and sublimation provided by language. Caught up in the horror, he cannot make of this experience a memory proper, in Freudian terms, which would mean its clear "cathexis to an object" available for representation and narration. '*La demente*' does not represent an appropriate object of the boys' sexual aggression because her provocative power overwhelms all efforts to satisfy these drives, which are felt to be 'always already' there, since they have been born into a social world in which these kinds of acts are permissible, and, indeed, necessary for the maintenance of social dominance. In this sense, these drives appear to lie beyond any form of representation—what could possibly bind or contain them? The dynamic possibilities of language have completely shut down. For these reasons, the boys' violent impulses leave them devoid of the psychological resource to cathect to objects in the 'here and now'. This is not even to mention the much more crippling psychological effects suffered by those who are victimized by 'white' male violence.

The 'patio interior' is a space of temporal immobility and sterility that coincides with the social and psychological inability of developing young males to gain maturity and understanding of the ways in which they are harming themselves and others. Indeed, there can be no 'healthy' coming-of-age when social belonging appears to depend upon sadistic violence that traumatizes and dehumanizes, above all, the victims, but also the perpetrators. It would seem that only the boys' increased unconsciousness of what they do would serve to alleviate their pain—a reversal of Freud's injunction to make the unconscious conscious—but this certainly would not make for a more just and humane world, and it would certainly not seem to indicate the true maturation of young boys into young men.

In contrast, Arguedas' evocation of the myth of *Inkarri* occurs in a particular context in which an aesthetic object of collective desire, the *zumbayllu*, exhibits its full dialectical power to change and be changed in positive ways that benefit the community, but only when the social actors who unleash its powers take a conscious and active role in the process. In contrast to assessments of the *zumbayllu*'s potential to effect personal and collective transformations that appear, at times, to rest on the limitations of language itself—Tarica's analysis of 'inner' *mestizaje* in which the Quechua language plays an inert and diffuse role similar to that of Saussurean *langue*, or Cornejo Polar's pessimism regarding linguistic and cultural hybridity—Arguedas suggests something very different.

Conclusion

If we return to Voloshinov's views of how language and other "ideological signs" change over time, we note commonalities to Quechua thought. Similar to the vibrant intersubjective relationships that form through ritual, aesthetic practices, Voloshinov writes: "outer ideological signs of whatever kind are engulfed in and washed over by the inner-signs" of individual psyches. This holds true even for larger linguistic grammatical forms—not just utterances—which from the standpoint of Saussurean linguistics comprise a synchronic system that "does not correspond to any real moment in the historical process of becoming" (66). In contrast to Saussure, Voloshinov writes, "what is important for the speaker about a linguistic form is not that it is a stable and always self-equivalent signal, but that it is an always changeable and adaptable sign" (68). There is no strict differentiation between grammatical forms (*langue*) and speech (*parole*); it is only one of degree. There is little about language per se that inhibits ideological shifts.

What, then, are the forces that inhibit positive ideological changes on a personal and collective level? The *zumbayllu*'s powers are at times drained, and it does, at times, seem to foment discord. Further, as we discussed in an earlier section of the essay, the boys appear to be unable to resist perpetrating violence against a perceived social inferior, despite the pain that it causes them, because their very identities depend upon domination of '*la demente*'. However, rather than consider the *zumbayllu*'s failure as related to immovable 'truths' about language and/or cultural mixing, and rather than view the grotesque nature of the boys' psychological development to be universally applicable, *Deep Rivers* suggests the need to historicize the workings of language and aesthetics. Cultural mixing can indeed function as a disempowering and hegemonic appropriation of an oppressed group's symbols, but this should come to mean that aesthetic creation cannot, by definition, facilitate relationships of intersubjective solidarity across diverse collectivities. Nor should we believe that by definition, young males must learn to repress their awareness of the domination that props toxic masculine identities. On the contrary, Arguedas encourages us to view the flimsy nature of these identities as reinforced by social structures that are ripe for upheaval and change.

Arguedas' answer to the question, "What are people for?", relates to the role language and aesthetics might play in building a more just world. *Deep Rivers* features the way in which certain aesthetic forms can connect us to the 'then' in the 'now' such that what seems to have been historically foreclosed becomes available to the imagination of a possible future. Arguedas reveals the stunning existence of the upside-down world of *Inkarri*, residing in the beautiful whirring of a child's toy as expressed through a language, which pulls together strands of a formerly alienated humanity. Of course, a child's toy alone cannot change the world. Just as the Vichian tradition of attentiveness to the historical nature of language and aesthetics inspired a generation of communists between the wars to struggle directly, Arguedas points towards the need to give history a push from more than one direction.

Notes

1 Frank Salomon defines the *ayllu* as "a landholding [Quechua indigenous] collectivity self-defined in kinship terms, including lineages but not globally defined as unilineal and frequently forming part of a multi-*ayllu* settlement" (22). Generally speaking, at the time the novel takes place, *ayllu* members enjoyed greater autonomy and were less vulnerable to the abuses of landowners than were other indigenous people.
2 MacCannell comes to refer to Freud's id as "It" in order to make an important distinction. Whereas Freud argues that the id contains repressed aggression against the father, MacCannell argues that the modern psyche is structured by unconscious, sadistic aggression against perceived 'others'.

Works cited

Arguedas, José María. *Deep Rivers*. Translated by Francis Homing Barraclough, University of Texas Press, 2002.
Bhabha, Homi. *The Location of Culture*. Routledge, 1994.
Brennan, Timothy. *Borrowed Light: Vico, Hegel, and the Colonies*. Stanford UP, 2014.
Cornejo Polar, Antonio. *Los Universos Narrativos de José María Arguedas*. Losada, 1973.
—. "Mestizaje, Transculturation, Heterogeneity." *The Latin American Cultural Studies Reader*, edited by Ana Del Sarto, Alicia Ríos, and Abril Trigo, Duke, 2004, pp. 116–120.
Fanon, Frantz. *Black Skin, White Masks*. Translated by Charles Lam Markmann, Grove Press, 1967.
Flores Galindo, Alberto. *Buscando un Inca*. Instituto de Apoyo Agrario, 1988.
Freud, Sigmund. *The Freud Reader*. Edited by Peter Gay, Norton and Company, 1989.

García, María Elena. *Making Indigenous Citizens. Identity, Development, and Multicultural Activism in Peru.* Stanford UP, 2005.

Herzog, Dagmar. *Cold War Freud.* Cambridge UP, 2017.

Larsen, Neil. *Determinations: Essays on Theory, Narrative and Nation in the Americas.* Verso, 2001.

MacCannell, Juliet Flower. *The Regime of the Brothers.* Routledge, 1991.

Rama, Angel. *Transculturación narrativa en América latina.* Siglo veintiuno, 1982.

Salomon, Frank. *The Huarochirí Manuscript: A Testament of Ancient and Colonial Andean Religion.* University of Texas Press, 1991.

Tarica, Estelle. *The Inner Life of Mestizo Nationalism.* University of Minnesota Press, 2008.

Taussig, Michael. *Mimesis and Alterity.* Routledge, 1993.

Uzendoski, Michael, Mark Hertica, and Edith Calapucha Tapuy. "The Phenomenology of Perspectivism: Aesthetics, Sound, and Power in Women's Songs from Amazonian Ecuador." *Current Anthropology*, vol. 46, 2005, pp. 656–662.

Vilaça, Aparecida. "Making Kin Out of Others in Amazonia." *Journal of the Royal Anthropological Institute*, vol. 8, 2002, 347–365.

Viveiros De Castro, Eduardo. "Cosmological Deixis and Amerindian Perspectivism." *Journal of the Royal Anthropological Institute*, vol. 4, no. 3, 1988, pp. 469–488.

Voloshinov, V.N. *Marxism and the Philosophy of Language.* Harvard UP, 1986.

11 Übermenschen and Untermenschen
Global Nietzsche and postcolonial fiction

Benjamin Noys

It is a trivial point to note that Nietzsche is a polarizing figure. Nietzsche's own self-presentation constantly stressed his divisive power and the necessity that his work create enemies. In *Ecce Homo* Nietzsche announced, "I am dynamite" (126), and in *Thus Spake Zarathustra*, Zarathustra, in a meditation on the friend, demands, "At least be my enemy!" (82). The history of Nietzsche's reception seems to bear out Nietzsche's own prophetic claims, with that history comprising a series of violent reactions, from outright condemnation to ecstatic celebration and identification. Sometimes, as in the case of Georges Bataille, these reactions are found in the same reader. While in the 1920s Bataille would powerfully criticize Nietzsche's "Icarian adventure" and celebration of the over-man as a "regression" (37), in the 1940s Bataille would write his *On Nietzsche*, in a deliberate act of mimetic identification. Nietzsche, therefore, at least succeeded in the task of division, scandal, and provocation, even as that has resulted in wildly divergent readings and the multiplication of 'Nietzsches', old and new.

In postcolonial studies, the response to the divisive problem of Nietzsche has been surprisingly muted. This is largely because thinkers like Derrida, Deleuze, and Foucault have mediated the appropriation of Nietzsche. The Nietzsche of postcolonial studies has been the Nietzsche of the philosophy of difference and of the critique of Western metaphysics, which is to say a Nietzsche shorn of his most disturbing elements. This postcolonial Nietzsche is part of what Neil Larsen referred to as the "new irrationalist hybrid" (#11). Operating within a philosophy of difference, Nietzsche has rarely been an explicit reference. We should note, however, Timothy Brennan's point: "Often posing as a single, coherent, philosophical and political ethics, postcolonial studies is really an uneasy mix of multiple schools of thought, criss-crossed by particular combinations of methodological differences, identitarian points of departure, regional foci, and political allegiances" ('Letters' 31). This "uneasy mix" certainly includes Nietzsche, but he has rarely played a direct role in the construction of the differentiation of positions within the field.

Recently, however, Nietzsche's polarizing effect has been activated in postcolonial studies, if from the margins. Timothy Brennan's *Borrowed Light* has drawn on Nietzsche's work to critically assess what we might call the 'colonial Nietzsche'. While Brennan notes Nietzsche is often assumed to be "the scourge of Europe" (*Borrowed Light* 133), in fact Nietzsche was enmeshed in reactionary and colonial politics. Far from the image of Nietzsche as the writer of multiplicity, Brennan traces Nietzsche's consistent aim of creating a new elite, worthy of the demands of European civilization. Nietzsche's "dissimulating style" disguises his "novel fantasy of conquest and European triumphalism" (*Borrowed Light* 195). In contrast, the thought of 'decoloniality', which activates what Walter Mignolo calls the "decolonial option" (161), has placed philosophical questions and problems as central, including the work of Nietzsche. Emerging from Latin America, decoloniality has insisted on the intertwining of colonialism with the colonial gestures of philosophy, although it also retains a tendency to articulate this in Heideggerean terms of "Western metaphysics." In the case of Nietzsche, Federico Luisetti suggests that it is the "Orientalism" of Nietzsche and his "naturalism" which offer the scope to rupture the transcendental regime of Western thought. Nietzsche's "political Orientalism" articulates a topology of life, which, Luisetti argues, is immanent and disruptive to the life-denying tendencies of Western thought (7). It is this turning to the outside, no matter how fantasmatic, which opens up the 'Western' to disruptive forms of life that cannot settle within a metaphysics that relies on the mechanical and the materialist to deny certain forms of life. What Brennan regards as the dissimulating image of Nietzsche as "the scourge of Europe," Luisetti takes as the true image of Nietzsche.

My aim is not to split the difference or mediate between these positions. Here, I aim to do something slightly different. Brennan has noted that Nietzschean thinking "labors under the impersonation of the novel" (*Borrowed Light* 218). Therefore, the actual novel might come into view as a site in which to probe the claimed mobility and evasiveness of Nietzsche's thinking.[1] Instead of the colonial Nietzsche or the decolonial Nietzsche, the Nietzsche I aim to take the measure of is the 'global Nietzsche'. The 'global Nietzsche' refers to the worldwide influence of Nietzsche, through a number of novels that appropriate and embody Nietzschean thinking in peripheral, colonial, and postcolonial contexts. In these cases, I want to suggest, the novel offers an 'impersonation' of Nietzsche, especially his thinking of the over-man, via the protagonists and plotting of the postcolonial novel. What is striking is how this 'impersonation' engages with Nietzsche's racialized politics of the 'over-man' to try to engage and reverse the position of the 'Untermenschen'. Donna V. Jones, analysing vitalism, notes how "an irrationalist, racially biologistic, eugenicist, and counterrevolutionary philosophical school—the very school that informed imperialist self-understanding—would appeal to colonial intellectuals seeking the rebirth of their cultures" (61). In a similar fashion, I want to probe these novels as they engage with the colonial politics of Nietzsche and attempt to reverse this politics. Such a gesture is one fraught with tensions and revealing of the problematic role of Nietzsche as thinker. While Brennan locates Nietzsche within the colonial moment of the late nineteenth and early twentieth century, to stress the colonial context of Nietzsche's philosophy, my complementary aim is to stress the dispersion of Nietzschean ideas into colonial and postcolonial contexts. What I will argue, in a development of Brennan's suggestion, is that we can see a new reading of Nietzsche that is developed from the periphery that questions Nietzsche's colonial politics, but also develops a new literary response to Nietzsche that reverses this politics.

I have selected a few novels that range geographically, to take some measure of the global Nietzsche. The first is Roberto Arlt's novel *Seven Madmen* (1929), an Argentinian novel that explores the ideologies of its 'madmen' plotting revolution in 1920s Buenos Aires. While not explicitly postcolonial, this work emerges from what Brennan calls the "global periphery" (*Borrowed Light* 139), and explores the ambiguous politics of 'revolution'. The second novel is Waguih Ghali's *Beer in the Snooker Club* (1964), which concerns Egyptian migrants to Britain in the early 1960s who remain torn between Nasser's 'revolution' and their own Anglophilia. What is striking about the novel is its reflections on the political form of this experience, in regards to the Left in Britain and Egypt, and the fates of its protagonists in relation to 'organized' politics. Finally, I examine Sam Selvon's *Moses Ascending* (1975), which follows the experience of Caribbean migrants who settled in Britain in the 1950s, in the figure of its protagonist Moses Aleotta, and their experience of the politicized 1970s. The novel is bitterly comic about the shortcomings of 'Black Power' movements, but also ironic about its own 'hero'. All three novels are broadly 'realist' in form, although displaying a realism that operates under pressure. We might speak, after the Warwick Research Collective (WReC), of a "peripheral realism," of novels that are "irrealist," particularly in their use of "un-rounded characters" (51). These peripheral texts register the fractures of combined and uneven development, but these formal fractures also interrogate the tensions between individual self-determination, in a Nietzschean style, and the contexts of projected or actual projects of collective self-determination, particularly communism. These novels belong to those "counter-current[s]" that incorporate "foreign forms" (WReC 56), in this case the "form" of Nietzschean philosophy and its imperial coordinates. These novel's displacements of Nietzsche reveal the ways in which 'Nietzscheanism' remains in a relation of dependence on collective social movements, even as it excoriates them. This revelation opens a critical space to re-read the Nietzschean 'over-man', as that over-man is written from the perspective, for Nietzsche, of the 'under-man'.

These are novels by men and, often, striking in their misogyny. This, however, is important to my analysis of their 'impersonation' of Nietzsche and the complex politics of gender at work in Nietzsche's texts. The politics at work in these novels are not 'pure', but fractured around colonial displacements, racism, the movements of the Left, and gender. Gillian Rose suggests, invoking Zarathustra visiting a woman with his whip, that "it is at the crux of gender that philosophy—the love of wisdom—the republic, and the legal fictions of the personality explode" (6). This scene in Nietzsche, of Zarathustra visiting a woman with his whip, is complex as Zarathustra receives this advice to bring the whip from an old woman. So, while misogynist this misogyny is passed off, by Nietzsche, onto women. It is, however, the very impersonation of Nietzsche, an impersonation of a philosopher impersonating a novelist, which allows us to reveal these tensions around gender and the colonial. The transplantation of Nietzsche into these very different global contexts, what I have called global Nietzsche, allows us to displace Nietzsche from his own position of mastery. If Nietzsche, according to Brennan, was master of a "politics of the literary" (*Borrowed Light* 136), then the "literary," with its own politics, might offer another way to continue the engagement with and critique of Nietzschean politics.

Astrologers of revolution

Seven Madmen is centred on the character of Remo Erdosian who, at the start of the novel, has been caught pilfering 600 pesos and seven cents from his employers. His wife then abandons him, and finally Erdosian plots to kill his wife's cousin. Erdosian lives in what he calls "the anguish zone" (*Seven* 6), which is "the product of man's suffering" and afflicts Erdosian with fantasies, both sexual and monetary, of exaltation and debasement. Erdosian is only one of the seven (perhaps more) madmen, who also include the Pharmacist, Erguet, with his religious mania and his claim to have discovered the "secret of roulette" (*Seven* 15). There is Arturo Haffner, 'The Melancholy Thug', a pimp, who once tried to commit suicide and is a violent misogynist, and Gregorio Barsut, Erdosian's wife's cousin. Bromberg, 'the Man who Saw the Midwife', and various candidates, including the Major and the Gold Prospector, complete the roster of madmen from the range of strange obsessives who inhabit Arlt's Buenos Aires. These fantasists inhabit a city that was, in the 1920s, the capital of one of the most prosperous nations in the world. Flooded with newly arriving European migrants, Buenos Aries offered a striking mix of prosperity and poverty; it was also flooded with radical ideas from Europe. Arlt's novel strikingly captures this situation of "combined and uneven development," both economic and cultural, in relation to the dynamic of the "capitalist world-system" (Shapiro and Barnard 29–44).[2] Buenos Aries is an archetypal "uneven city" (WReC 143–167).

Arlt's novel is heavily indebted to Dostoevsky and to his "hallucinatory realism" (WReC 61; Caistor 309). His own work as a journalist and crime reporter attunes his novel to the diversity of types in the modern city—he wrote a column called *Buenos Aries Sketches*. The novel's impulse to realist sketches of various characters is also engaged with the complexity of the metropolitan space of the "uneven city," which generates Arlt's "irrealism." Arlt's own struggle with style—he thought he wrote "badly," due to economic pressure—is another sign of this struggle to formally engage with a sudden and rapid social development (Caistor 312). This is why *The Seven Madmen* is a novel of pressures: the pressures which make the madmen mad, the pressure of the sudden irruption of development in the periphery, and the pressure of social, political, and literary forms that creates an unstable desire for revolution.

Marx and Engels famously made a critique of those who engaged in conspiratorial versions of revolution, the Blanquists, as "alchemists of the revolution," who "are characterized by exactly the same chaotic thinking and blinkered obsessions as the alchemists of old" (Marx and Engels 318). In Arlt's *Seven Madmen* it is 'the Astrologer' (*el Astrólogo*) who is at the centre of similar revolutionary schemes. Like the Blanquists, who "leap at inventions which are supposed to produce revolutionary miracles" (Marx and Engels 318), Arlt's Astrologer and his other madmen try to 'invent' the revolution as a kind of machine. They are tinkerers (*bricoleurs*) of revolution, who are obsessed with devices, schemes, and machines. The Pharmacist says, "Who is going to make the social revolution if it's not the swindlers, the wretched, the murderers, all the scum that suffer here below without the slightest sign of hope?" (*Seven* 18). Again, like the Blanquists, this is a plebeian or lumpen version of revolution.

Erdosian suspects the Astrologer "might be a Bolshevik agent" (*Seven* 30). The Astrologer, in fact, takes 'revolutionary' inspiration from a range of sources, including the Ku Klux Klan, and suggests that: "I don't know if our society would be Bolshevik or fascist. Sometimes I think the best thing would be to concoct such an unholy mixture that not even God could untangle it" (*Seven* 34). The Astrologer's secret society is a "super-modern one" (*Seven* 34), in which members would contribute money as if in a bizarre revolutionary pyramid scheme. Its other source of finance is brothels, playing once again on Arlt's mixing of sexual and political fantasy. The leaders in this revolution will be the great industrialists, and the Astrologer aims "to make industry mystical" (*Seven* 40). These industrialists belong to the "realm of supermen" (*Seven* 152), but the immense power they hold can crush any revolution. In its place, the Astrologer dreams of a secret society that will ruthlessly exploit the young, and in what is a parody of capitalist development the society will force workers into new industrial schemes. They will even create their own deity, in the form of an attractive youth, who will lure people into this 'revolution'.

To finance this mad revolution the Astrologer agrees to Erdosian's plan to kidnap Barsut, who had reported Erdosian to his employers as he was in love with Erdosian's wife, and extort money from him by torture. The conspirators meet together and one, the Major, suggests the ripeness of the army for a military coup and dictatorship. This was to happen in Argentina in 1930 with a coup led by José Félix Uriburu, a fact noted by the commentator (a narrative voice in the novel who comments on events in footnotes). The Major aims to create a fictitious revolutionary force, arrange terrorist attacks, and try to provoke "all the dark, ferocious forces in society" (*Seven* 175). The military will then step in to bring order and dictatorship. It turns out, however, that the Major is not in the military, but a plant by the Astrologer. Still, the inchoate desires of the madmen, which all seem to focus on a "strong man," whether that be Lenin or Mussolini, and the realization of wealth and power, remain eerily prescient. In fact, this milieu of inventors and tinkerers, sexually frustrated and wanting revenge on society, echo, today, the 'Alt-right' and its technological fascisms.[3]

The Astrologer and Erdosian eventually contemplate the murder of Barsut to get his money. Once again, the Astrologer is obsessed with images of strength, power, and violence. He is fascinated by Lenin who said, "This is madness. How can we make a revolution if we don't shoot anyone?" (*Seven* 264). Barsut hands over the money to finance the scheme, but the Astrologer and Erdosian still plan to kill him, with the deed to be done by Bomberg, the Man who Saw the Midwife.

In one final twist, however, unknown to Erdosian, in the end the Astrologer and the Man who Saw the Midwife fake the murder. After the Dostoevskian anguish of Erdosian we are delivered one final fiction that reduces that anguish to farce. The claim to realism has broken down in an 'irrealism' that ruptures the Dostoevskian desire to murder. There is no moment of authenticity, only another fiction that leaves the novel unresolved and Erdosian as a truly un-rounded character.

Arlt's fiction undermines the pathos that Nietzsche invests in the figure of the 'madman'. The most famous instance of this investment is the announcement of the death of God, in Nietzsche's *The Gay Science*. The madman is searching for God and announces to the mocking crowd that "*We have killed him*—you and I" (*Gay Science* 119–120). The world is unmoored from its orientation and we even "smell nothing of the divine decomposition" (*Gay Science* 120). The madman realizes his announcement is too early and sings the funeral service for God in several churches. There is also, of course, the pathos attached to Nietzsche's own 'madness', and his own announcement, "every name in history is I" (*Selected Letters* 347). While Nietzsche's pathos of madness is deliberately 'world-historic', in the case of Arlt's madmen world-history is eyed from the periphery. This is most evident in the speech by the Astrologer, which is a virtual parody of Nietzsche's madman. The Astrologer declares that our loss of faith will lead to a plague of suicides, and that humanity suffers a "dreadful metaphysical sickness" (*Seven* 154), which requires purging by violence to reveal the supermen. The new leaders, of which he will be one, will "terrorize the weak and arouse the strong" and "exalt barbarity" (*Seven* 163).

Robert S. Wells argues that we see Arlt as a reader of Nietzsche,[4] and the Astrologer as Arlt's most Nietzschean character (n.p.). It is the Astrologer who proclaims, "Many of us have a superman inside. The superman is our will when it is fully realized" (*Seven* 296). Yet, as Wells also points out, the Astrologer "makes a monster out of Nietzsche" (n.p.). While Wells stresses the discordance and disorder this introduces into our understanding of Arlt and Nietzsche, we can add that this rendering of Nietzsche as a 'monster' also reveals the authoritarian and 'ordered' politics at work in Nietzsche. The Astrologer's version of Nietzsche's "I am all the names in history" is that the leaders of his secret society will "be Bolsheviks, Catholics, fascists, atheists or militarists, depending on the level of initiation" (*Seven* 161). The Astrologer, like Nietzsche, wants to introduce a 'caste' separation between leaders and the masses. The explosion of identity serves an order and hierarchy of castes, which Arlt's madmen try to subvert by seizing the dominant position. The Astrologer's aim to re-invent religion and myth can be read as not so much a betrayal of Nietzsche's attempt to go "beyond good and evil" as the recognition of Nietzsche's own reactionary politics.

The Astrologer says that for revolution, "All you need is willpower and money" (*Seven* 274). Arlt's novel reveals not only the authoritarianism of Nietzschean politics, but also the role of money that is usually hidden in invocations of the will-to-power. If the will-to-power is inflated to a metaphysics that claims to overcome all limits, in Arlt's novel it always encounters the limit of money. Ricardo Piglia has pointed out the centrality of money to Arlt's fiction, as a "machine for producing fictions" (13). In the case of *Seven Madmen* the secret society constructed by the Astrologer is "both a story-producing and money-producing industry" (Piglia 13). Alchemy is not the alchemy of revolution, but the alchemy of making money. Erdosian, with his inventions, tries to make money out of nothing. According to Piglia, "[h]is inventions . . . are a sublimated, alchemic form of capitalist processing in which it is not a question of working with concrete goods but with ideas of goods, essences of money" (15). Piglia concludes that Arlt identifies writing with the power to make money. It is the ability to make fictions that can engage with the fiction of money and Arlt's novels are another form of machine, although ones that break down, misfire, stutter, and require constant tinkering to function.

Nietzsche at the periphery inhabits a different perspective in Arlt's novel. Arlt's characters live, in fact, in one of the cities that are central to the world economy, but they live that experience as one of alienation and exclusion. Their delirium is indicative of a very different world-historical crisis to that of 'European nihilism', as they seek dominance through schemes to claim their 'rightful' place as leaders, as supermen. The madmen reveal the gendered and class-based fantasies that drive the concept of the 'superman' in ways that postwar readings of Nietzsche in Europe tend to deny or minimize. Arlt's peripheral reading, conducted in the 1920s, is more revealing of those concurrent readings of Nietzsche that stressed his authoritarianism and colonial politics. His fiction is different in that rather than trying to reinforce these fantasies to drive a fascist or Nazi politics, his fiction explodes these fantasies and so reveals Nietzsche's political and metaphysical investments.

Revolution and *ressentiment*

Waguih Ghali's only novel, *Beer in the Snooker Club* (1964), published before his suicide in 1969,[5] is a reflection on the experience of displacement and exile, between Egypt and London, at the time of Nasser's 'revolution'. The novel's anti-hero is Ram, a wastrel scion of the Coptic elite in Egypt, and the novel is focused on his relationship with his friend Font. Both characters are torn between Anglophilia, a desire to support a more radical revolution in Egypt, and the contradictions of their own class positions. Ram describes Font as "about the only angry young man in Egypt" (*Beer* 15), and Font comments that "we're so English it is nauseating" (*Beer* 18). The novel begins in Egypt, after Ram and Font have returned from a four-year stay in London. Ram is living a parasitic life where he has "drifted on that rich tide" of the wealth of his extended family (*Beer* 24), while securing for Font a job in the snooker club. The ritual they have in the club is of pouring Egyptian beer with some vodka and whiskey to create something that is close to the Draught Bass beer they drank in Britain. These drinks are mixed in two silver tankards that Ram and Font bought in London and had engraved with their names.

This nostalgic ritual, both pathetic and meaningful, speaks as an obvious metaphor for Ram and Font's divided consciousness. The novel uses a realist form to explore this division and is episodic and drifting in character, mimicking its characters' movements between Britain and Egypt and their unmoored lives. Ram and Font are not heroic Nietzschean supermen, but characters caught between differing failures of Left politics in Britain and in Egypt. Ram says he has known many "Fonts," "who are not revolutionaries or leaders in the class struggle, but polished products of the English 'Left', lonely and without luster in the budding revolution of the Arab world" (*Beer* 20). For Ram the problem is of the relentless obsession with Britain and the detachment from Egypt, with his friends using "English personalities as a nucleus" for their political discussions that "circle round and round" (*Beer* 34). While Ram may not be an obvious 'superman', he does express a jaded disillusionment with the impotence of politics, although one reflective of the problems of the Left.

Six years earlier Ram had been involved in protests against the British troops and Suez and is seen as a 'red' by his relatives. This is the period of his political education when "the only important thing that happened to us was the Egyptian revolution" (*Beer* 52). Ram begins a sexual relationship with Edna, who is Jewish and the daughter of a wealthy family. She is also, however, a communist, and mingles with the ordinary people that Ram remains distant from. Edna also finances Ram and Font's trip to England. Ram experiences England as an education, but is sceptical about the results. Ram claims that the "mental sophistication of Europe" has caused him to lose his "natural self" and that he has become a "fictitious character" (*Beer* 60), and this experience of alienation is central to the novel. While 'existential', this experience is not Nietzschean. The Nietzschean approach would suggest the embracing of the fictional as the true state of the world became a "fable" (*Twilight* 40–42). Instead, Ram and Font are obsessed with the risk of 'phoniness' while in England. Edna, however, is doubtful as to Ram's claims to authenticity as an Egyptian, due to his class position. She says, "You are what you are; and that is a human being who was born in Egypt, who went to English public school, who has read a lot of books, and who has an imagination" (*Beer* 106). Ram's experience in England is one shaped by rage and dissolute behaviour, and this sense of inauthenticity.

The novel returns to Egypt and Ram's relationship with Edna, which is one of "torturing pressures" (*Beer* 122). Ram is working for Dr Hamza, who reports the atrocities of the Nasser regime, by collecting photographs of torture from police officers for money. Ram's decision to send the photographs to a series of newspaper editors endangers the operation and so his new political gesture ends in failure. He also admits to Edna that while in England he had joined the Communist Party. The reason Ram gives for joining was that his politics was mixed up with his love for Edna and, as a result, he was left with a knowledge of suffering that he did not know what to do with. Ram says that if someone has imagination and intelligence then "there are two things that can happened to him; he can join the Communist Party and then leave it, wallowing in its shortcomings, or he can become mad" (*Beer* 189). Yet, Ram has left the Party and appears on the edge of madness. He is painfully aware of the rift in his own experience: "I see myself not only through Egyptian eyes, but through eyes which embrace the whole world in their gaze" (*Beer* 191). Instead of a colonial hybridity that lies between two identities, Egyptian and English, Ram experiences also a gaze that embraces the world—communist internationalism. This 'world' is an intolerable burden, precisely because its failures leave Ram without a resolution to his divided experience.

The resolution of the novel is suitably low key. Ram decides to marry Didi Nackla, a wealthy young Egyptian woman he had slept with in London. Her two conditions are his relationship with Edna is resolved, hence Ram tells her Edna is already married, and that Ram give up "this political business" (*Beer* 220). While telling his fiancée he is off to tell his political friends he is no longer interested, Ram secretly decides to go out gambling. I have stressed that the novel appears as un-Nietzschean, but while not offering a direct impersonation of Nietzsche it does put Nietzschean politics under pressure. Ram's sense of phoniness and his divided identity are not celebrated as instances of fiction or hybridity,[6] but instead they are grounded in his class and 'national' position.

In addition, the novel explores the tensions of revolution and the failures of the revolutionary process in ways that dispute Nietzsche's claim that all forms of 'Left' radicalism are instances of *ressentiment*. Nietzsche, especially in *On the Genealogy of Morals*, will argue that reactive "slave revolt" is merely an act of envy, an inability to inhabit the will-to-power, and the destructive sign of the agitator and socialist. While Ram's wealthy relatives constantly regard his 'communism' as infantile and reactive, implicitly adopting a Nietzschean diagnosis, Ram experiences revolution and its failures as a passion that is not completed. In this sense his desire is not so much 'revenge', but to see a revolution in which there would no longer be a tide of wealth he could coast on and in which "all tides had vanished" (*Beer* 24). Contrary to Nietzsche's claim that *ressentiment* reinforces a reactive identity—slave versus master or communist versus capitalist—Ram remains true to Marx's insight that a revolution would be the dissolution of class society (if driven by those who have nothing within that society).

Fredric Jameson has noted the paradox of the Nietzschean diagnosis of *ressentiment*, which is that to diagnose *ressentiment* is to express resentment (*Political* 189). Jameson argues that the analysis of *ressentiment* is "little more than an expression of annoyance at seemingly gratuitous lower-class agitation" (*Political* 189). He also argues that "*Ressentiment* is the primal class passion" ('Marx's Purloined Letter' 39). Certainly, Ram experiences that passion and experiences the failure of that passion to shift from a 'primal' position to one that is collective and organized. *Beer in the Snooker Club* is not only a novel of the failure and compromise of Ram with his class 'destiny', but also of the failure of the Left in England and Egypt. The drifting form of the novel signals this failed resolution, as it passes backward and forward in time and across the space between metropole and periphery. Ossified communist formations become the only choice to avoid madness, but remain unliveable as mechanisms that could produce the revolution that would resolve the class fragmentation Ram experiences. Instead of the blanket psychologization of revolution that Nietzsche presents, to reactionary ends, the novel undermines the psychological by tracing its fragmentation through contradictory class and colonial relations.

Black Power

We first encounter Moses Aloetta, the central character of Sam Selvon's *Moses Ascending*, in Selvon's 1956 chronicle of West Indian migrant experience, *The Lonely Londoners*. In *The Lonely Londoners* Moses is self-described as a "liaison officer" and "welfare officer" to "the boys," his fellow arrivals from the Caribbean (*Lonely* 2, 3). Most of the novel is concerned with telling the travails of the various "boys" in the form of a series of "ballads," comic tales or stories that echo Trinidadian calypso and which try to capture the various types inhabiting the imperial metropole. Late in the novel organized politics intrudes. Moses has a conversation with his friend Galahad in which Moses bitterly reflects on his ten years of living in this "lonely miserable city" (*Lonely* 126). The optimistic Galahad suggests Moses needs a holiday, a trip to Berlin or Moscow, as "I hear the Party is giving free trips to the boys to go to different cities on the continent, with no strings attached, you don't have to join up or anything" (*Lonely* 127). Moses is sceptical and finally cautious: "Take it easy, I have enough grey hairs as it is" (*Lonely* 127). Whatever the decision, the Party is not represented as a serious political choice, but merely the opportunity for travel.

While Moses is cautious, he does go on to say he campaigns for the Labour Party at election times and to express doubts about their entrepreneurial friend, Harris: "He tell me Labour, but I have a mind he is a Tory at heart. He always taking about the greatness of the old Churchill and how if it wasn't for him this country go right down" (*Lonely* 129–130). If sceptical about organized politics, Moses functions as this fulcrum of a new community, in the process of forming in relation to British racism. The light-skinned Bart on arrival tries to pass as Latin American, but in the face of the "old diplomacy" of the British, "Bart boil down and come like one of the boys" (*Lonely* 48). Every Sunday "the boys liming in Moses room" (*Lonely* 134), and Moses reflects that:

> Sometimes during the week, when he come home and he can't sleep, is as if he hearing the voices in the room, all the moaning and groaning and sighing and crying, and he open his eyes expecting to see the boys sitting around.
>
> (*Lonely* 135)

In a way Moses is the focus of this new collective subject, of the new community of the Caribbean "boys" and its often unhappy point of convergence—in the same way, as Susheila Nasta has argued, the novel itself forms a "community of words."

Selvon's 1975 sequel *Moses Ascending*, set twenty years later, explores the tensions and collapse of this fragile collective subject in the face of ongoing racism and the new politicizations of Black Power. These tensions are re-staged between Moses and Galahad, between Moses who is now a landlord owning a ramshackle property and Galahad, who is now a militant of Black Power. Moses has decided to abandon his past friends for the sake of his own individual success and has no time for Black Power: "I didn't have anything to do with black power, nor white power, nor any fucking power but my own" (*Moses* 18). This credo might well correspond to what Ishay Landa has called "the overman in the marketplace," in which the Nietzschean will-to-power is correlated with the power of capitalist success. Moses has fully adopted an individualism that desperately clings to power and has "internalized the values of capitalist, white Britain" (Dyer 130). Moses himself recognizes the change: "the rumor went around town that I was a different man, that I had forsaken my friends, and that there was no more pigfoot and peas and rice, nor even a cuppa to be obtained" (*Moses* 5). This parodic and carnivalesque sense of reversal is perhaps best represented by Moses' 'manservant' Bob. Moses adopts Bob, a white man come to London from the Midlands who cannot afford to pay rent, in an explicit parody and reversal of Crusoe's relationship with Friday (Kunzru; Rampaul).

The novel is more conventional in form than *The Lonely Londoners*, tracing a comic logic of Moses 'ascending' and 'descending' as he tries to inhabit the values of the "overman in the marketplace." Much of the effective comedy of the book comes about because while Moses claims to have become a radical individualist and capitalist, his intentions constantly go awry. We could call this Hegelian comedy, in the sense of Hegel's notion of the "cunning of reason," carried out not at the level of history but at the level of the individual. While Moses constantly lambasts the Black Power Party he ends up, accidentally, attending their demonstration and finds himself arrested. He will later end up hosting the Party headquarters in his basement, and every attempt at distancing himself only leaves him further involved. Similarly, Moses expresses his racist attitudes to the new immigrants from South Asia, whom he always refers to by the racist epithet 'Pakis', and he seems unable to distinguish between Muslims and Hindus. Yet, he becomes involved in a scheme of sheltering illegal immigrants, albeit for money, and ends up temporarily giving over his own penthouse flat to house them. Finally, due to his softheartedness to his newlywed 'manservant' Bob, he gives up his flat to Bob and his wife. Moses will descend to the basement and the ascension of Moses has proved comically short-lived.

Moses Ascending mixes "both a carnivalesque atmosphere and an underlying tone of cynicism" (Kunzru n.p.). Rebecca Dyer suggests we read Moses' descent as a satire of his unpleasant and self-serving views (139), but it is difficult to escape, as she concedes, the unflattering presentation of Black Power activists as self-serving and deluded. Moses' relationship with the young militant Brenda is characterized by his "unreconstructed sexism," although this is mixed with a qualified admiration for her ability to become a "Black Briton" (Kunzru n.p.). Galahad the militant is simply on the make: asking for money from Moses for the "struggle," happy for Moses to be arrested to create a *cause célèbre*, and using his basement for Party business. The Black Panther activist from the United States, who exhorts his London comrades to "kill all whites and burn down the City of London" (*Moses* 123), drives a Mercedes and absconds with the funds of the British Party. While intended as comic, the comedy relies on stereotypes and clichés that would become the province of the ascendant Right. The novel is not so far from the political common sense of Thatcherism, in which every 'alternative' political project is derided as both impotent and dangerous. Moses himself refuses the notion that black people could ever unite and embraces the individualism of his own 'power'.

Nonetheless, I want to suggest the novel raises the question of the conception of life and politics in terms of will-to-power, even against its own intentions. Certainly, Nietzsche was among the philosophical influences on the Black Panthers, with Huey Newton suggesting that while he did not endorse all of Nietzsche he did recognize that "Nietzsche was writing about concepts funda-mental to all men, and particularly about the meaning of power" (in Caygill 12). In a 1974 article, 'The Mind is Flesh', Newton would adopt and adapt Nietzsche's 'over-man' in the figure of "a new sort of man, capable of preserving, amplify-ing, and passing to our human or posthuman followers the striving for mastery of reality, while preserving its elements of intellect, character, freedom, and joy" (in Caygill 13). While this is a significant rewriting of Nietzsche, Selvon's novel reveals some of the risks of a 'Left Nietzscheanism'. If politics becomes a matter of competing wills-to-power then demands for equality lose their rational basis. While Moses might well be self-serving, his own recognition of politics being simply a matter of power implies a cynicism that suggests competing powers lack the ability to form real communities of resistance.

The novel's own carnival reversals lend it an instability that is oddly Nietzschean. These carnivalesque moments are not so much represented formally—as we have noted, the novel remains fairly linear—but emerge in the 'arc' of the text. More painfully, the formal carnival is also reflected in Moses' malapropisms and his own claims to literary writing—he is working on his own novel out of the material that makes Selvon's novel. We should not underestimate the cynicism that drives the narrative, which, as we have seen with Ghali's novel, can engage the Nietzschean theme of *ressentiment*. My aim is not to redeem *Moses Ascending*, but to suggest that the way in which it plays the problem of power and of reversal is revealing of the political instabilities contained in the attempts to deploy Nietzsche for radical ends. While the cynicism towards revolutionary politics and collectivity marks it as Nietzschean, the bathos of Moses' descent from penthouse to basement plays a role that undermines Nietzschean pathos. We could also say this extends to the deflation of Moses' literary ambitions, as he is by no means the stylist and master of language he supposed. The reversal of over-man to a literal under-man is a moment of the crashing of Nietzsche's 'Icarian' adventure. While this is displaced from Nietzsche's authoritarianism and colonial politics, this displace-ment also suggests the limits of attempts to detach a 'Left Nietzsche' from this authoritarian context.

Impersonating Nietzsche

Nietzsche is well known as the philosopher who celebrates the mask, as in *Beyond Good and Evil*, when he suggests "every profound spirit needs a mask" (51). Gilles Deleuze suggests that Nietzsche's philosophy is fundamentally theatri-cal, and that the multiplicity of Nietzsche's texts is one of different characters and masks donned by Nietzsche (*Difference* 5). Deleuze argues that the various historical personages that litter Nietzsche's texts "are designations of intensity" ('Nomad' 146), and it is precisely this dispersion that suggests those texts of Nietzsche's with fascist or anti-Semitic 'resonances' are merely one set of forces amongst others and the true force is a revolutionary one ('Nomad' 145, 146). Deleuze argues Nietzsche's capacity to impersonate various figures, to take on various masks, liberates those texts from political responsibility and explodes the limits of identity ('Nomad'). In reply, we could criticize, first, the monotony of Nietzsche's masks and impersonations, which remain, recognizably, Nietzsche. Second, we could note the evasiveness of this mode of reading, in which the historical forces and significations attached to various 'masks' are obliterated by the overwhelming force of the will-to-power.

My reading suggests a reversal and a displacement. No longer is it a matter of the theatrical, literary, and staged functions of Nietzsche's texts, but of Nietzsche being donned as a mask. Instead of the critical task of locating the historical significations attached to Nietzsche's masks, these impersonations of Nietzsche's displace his literary politics to the periphery. Reversal is, of course, disputed by Nietzsche as a mere act of *ressentiment* and this rejection has become the norm in the anti-dialectical arguments of contemporary French philosophy. I have suggested, however, that these displacements of Nietzsche to the global are not simply reversals. While they run the risk of being understood in terms of Nietzsche's concept of *ressentiment*, they also disrupt that coding by pushing against the limits of Nietzsche's aristocratism. In the case of each of the novels we have examined, the displacements between the over-man and the under-man have also suggested the fraught engagements with other collectivities and other political possibilities. All three novels can be treated as exercises in cynicism towards organized revolutionary politics, which is to say read in a Nietzschean fashion. My counter-reading has aimed to suggest the other possibilities of "eyes which embrace the whole world in their gaze," as Ghali puts it (191). This embrace of the world suggests a politics of internationalism and equality, a utopian horizon, which is also crucial to these novels and contrary to Nietzschean cynicism.

This may also account for the use of a 'peripheral realism' that, while touching on 'irrealism', does not fully embrace modernist fragmentation or the plurality Nietzsche claims as the mark of his style. The very awkwardness of style and form, especially in Arlt and Selvon, suggests the pressures of fragmentation and dispersion that are also contained within a desire to provide a narrative that can express collectivity. Instead of the reading that suggests fragmentation and dispersion open out to a multitude, instead of the Nietzschean mimicry of "all the names of history," here we have novels that put pressure on narrative and form while retaining the sense of the collective as a mode of self-determination. These "un-rounded characters," these obsessives and madmen, are not only under the pressure of capitalist modernity, but also under the pressure of revolutionary politics. The strains and stresses of literary realism at work engage with this problem, rather than dissolving it into plurality and fragmentation.

In each case, I have been keen to trace the tensions registered in these texts about these possibilities. In particular, we can note the particular tension that falls not only on the colonial but also on the relation to gender. Erdosian's 'black house' of masturbatory fantasy in *Seven Madmen*, Ram's serial womanizing, and Moses' sexual reveries on the availability of "black beauties" (*Moses*) all suggest the intertwining of misogyny with a Nietzschean politics of power. Without excusing these forms of misogyny, these texts at least reveal and engage with the dominant link between Nietzsche's repudiation of femininity and his colonial and anti-socialist politics. While Derrida recognizes Nietzsche's virulent misogyny, his reading of Nietzsche aims to affirm the "figure" of woman in Nietzsche's texts as a moment of "affirmative power," beyond truth and untruth (97). Such an affirmative escape from the problem of gender and truth leaves these effects of power behind for a utopia that transcends history and domination. My suggestion is the staging of these fantasies by literary subjectivities that are not supposed to participate in the Nietzschean project causes these forms to explode.

Malcolm Bull, in his *Anti-Nietzsche*, suggests we rewrite Nietzsche's texts from the perspective of the "losers": women, socialists, and all those expelled or demeaned in Nietzsche's demand for a "great politics." The novels we have explored do something similar, but also different. They rewrite Nietzsche by trying to inhabit Nietzschean themes, especially the will-to-power, from the position of the "losers." In this impersonation of Nietzsche the claims for the plurality and multiplicity of Nietzsche's texts, often deployed to disable critical reading, are placed under pressure. Nietzsche's "literary politics," which are supposed to remove rational control over meaning are, instead, exploded from within. This is not to argue that these texts have solved the problem of Nietzsche or rendered his texts or philosophical project non-toxic. Instead, rather, they push us further into the problems of Nietzsche's politics without providing us with a purified 'New Nietzsche'. The global Nietzsche, as I have called this strategy of reading, is not simply the name for the expansion of Nietzsche's influence over the world, but also the name for rendering that influence as a global problem. It is the particular peripheral position of these writers, their self-aware status as under-men, which offers traction on the Nietzschean vision.[7] At this point, as we have seen, other possibilities emerge that have been repressed and forgotten. The collective projects of self-determination that haunt these texts re-emerge as possibilities that challenge us to go beyond Nietzsche.[8]

Notes

1 See Nehamas, *Nietzsche* for a positive reading of Nietzsche as 'literary' philosopher.
2 See also WReC, *Combined and Uneven Development*.
3 See Nagle, *Kill All Normies* for an initial description of the culture of the 'Alt-right'.
4 In Arlt's novel *The Mad Toy* the central character's library includes Nietzsche's *Antichrist* (77), alongside a volume on electrical engineering.
5 See Athill, *After a Funeral*, for Ghali's life in London and the circumstances around his suicide.

6 See Homi Bhabha, *The Location of Culture*, for an implicitly Nietzschean celebration of hybridity as 'fiction' and subversion.
7 Comparison could be made with other, more 'central' writers, who also displace Nietzsche in various ways, such as H.G. Wells, George Bernard Shaw, D.H. Lawrence, and Jack London. I owe this point to Ishay Landa.
8 I would like to thank Fiona Price, Harrison Fluss, and Ishay Landa for their comments and criticisms.

Works cited

Arlt. Roberto. *The Mad Toy*. Translated by James Womack, Hesparus Press, 2013.
—. *The Seven Madmen*. Translated by Nick Caistor, Serpent's Tail, 2015.
Athill, Diana. *After a Funeral*. Granta, 1986.
Bataille, Georges. *Visions of Excess: Selected Writings 1927–1939*. Edited and introduced by Allan Stoekl. Translated by Allan Stoekl et al., U of Minnesota P, 1985.
Bhabha, Homi. *The Location of Culture*. Routledge, 1994.
Brennan, Timothy. *Borrowed Light: Vico, Hegel, and the Colonies*. Stanford UP, 2014.
—. "Letters from Tunisia: Darwish and the Palestinian State of Mind." *CounterText*, vol. 1, no. 1, 2015, pp. 20–37.
Bull, Malcolm. *Anti-Nietzsche*. Verso, 2011.
Caistor, Nick. "Arlt's Life and Times." In Roberto Arlt, *The Seven Madmen*, pp. 307–312.
Caygill, Howard. "Philosophy and the Black Panthers." *Radical Philosophy*, no. 179, 2013, pp. 7–13.
Deleuze, Gilles. *Difference and Repetition*. Translated by Paul Patton, The Athlone Press, 1994.
—. "Nomad Thought." *The New Nietzsche*, edited by David B. Allison, The MIT Press, 1985, pp. 142–149.
Derrida, Jacques. *Spurs: Nietzsche's Styles*. Translated by Barbara Harlow, Chicago UP, 1979.
Dyer, Rebecca. "Immigration, Postwar London, and the Politics of Everyday Life in Sam Selvon's Fiction." *Cultural Critique*, vol. 52, no. 1, 2002, pp. 108–144.
Ghali, Waguih. *Beer in the Snooker Club*. Serpent's Tail, 2010.
Jameson, Fredric. "Marx's Purloined Letter." *Ghostly Demarcations, A Symposium on Jacques Derrida's* Specters of Marx, edited by Michael Sprinker, Verso, 2008, pp. 26–67.
—. *The Political Unconscious*. Routledge, 2002.
Jones, Donna V. *The Racial Discourses of Life Philosophy: Négritude, Vitalism, and Modernity*. Columbia UP, 2012.
Kunzru, Hari. "Introduction." *Moses Ascending*. Kindle ed., Penguin, 2008, n.p.
Landa, Ishay. *The Overman in the Marketplace*. Lexington, 2007.
Larsen, Neil. "Postmodernism and Imperialism: Theory and Politics in Latin America." *Postmodern Culture*, vol. 1, no. 1, 1990, http://pmc.iath.virginia.edu/text-only/issue.990/larsen.990. Accessed July 11, 2017.
Luisetti, Federico. "Nietzsche's Orientalist Biopolitics." *BioPolítica*, 2011, www.biopolitica.cl/docs/publi_bio/luisetti_nietzsche.pdf. Accessed May 18, 2017.
Marx, Karl, and Friedrich Engels. *Collected Works, Volume 10*. Lawrence and Wishart, 2010.
Mignolo, Walter D. "Epistemic Disobedience, Independent Thought and Decolonial Freedom." Theory, Culture and Society, vol. 26, nos. 7–8, 2009, pp. 159–181, doi: https://doi.org/10.1177/0263276409349275.

Nagle, Angela. *Kill All Normies*. Winchester: Zero Books.

Nasta, Susheila. "Setting Up Home in a City of Words: Sam Selvon's London Novels." *Other Britain, Other British*, edited by A. Robert Lee, Pluto, 1995, pp. 48–68.

Nehamas, Alexander. *Nietzsche: Life as Literature*. Harvard UP, 1985.

Nietzsche, Friedrich. *Beyond Good and Evil*. Translated by R.J. Hollingdale, Penguin, 1973.

—. *Ecce Homo*. Translated by R.J. Hollingdale. Penguin, 1979.

—. *Selected Letters of Friedrich Nietzsche*. Edited and translated by Christopher Middleton, Hackett, 1996.

—. *The Gay Science*. Edited by Bernard Williams, Cambridge UP, 2001.

—. *Thus Spake Zarathustra*. Translated by R.J. Hollingdale, Penguin, 1961.

—. *Twilight of the Idols / The Anti-Christ*. Translated by R.J. Hollingdale, Penguin, 1968.

Piglia, Ricardo. "Roberto Arlt and the Fiction of Money." Translated by John Kraniauskas, *Journal of Latin American Cultural Studies*, vol. 8, no. 1, 1999, pp. 13–16.

Rampaul, Giselle. "A. Black Crusoe, White Friday: Carnivalesque Reversals in Samuel Selvon's *Moses Ascending* and Derek Walcott's *Pantomime*." *Culture, Language and Representation*, vol. 1, 2004, pp. 69–80.

Rose, Gillian. *Dialectic of Nihilism*. Basil Blackwell, 1984.

Selvon, Sam. *Moses Ascending*. Kindle ed., Penguin, 2008.

—. *The Lonely Londoners*. Penguin, 2006.

Shapiro, Stephen, and Philip Barnard. *Pentecostal Modernism: Lovecraft, Los Angeles, and World-Systems Culture*. Bloomsbury, 2017.

Wells, Robert S. "Orders and Disorders, Wills to Powers—Arlt and Nietzsche, the *Astrólogo* and the sociedad secreta." *Dissidences*, vol. 5, no. 9, 2013, Article 11, n.p., http://digitalcommons.bowdoin.edu/dissidences/vol5/iss9/11. Accessed June 20, 2017.

WReC (Warwick Research Collective). *Combined and Uneven Development: Towards a New Theory of World-Literature*. Liverpool UP, 2015.

Afterword

Timothy Brennan

What is the force of the word 'poetic' in this volume? To what is it referring exactly? It does not seem to be *poiesis*—a popular term in theory that we associate most with Hannah Arendt, who set out to crack the codes of contemporary political life by identifying what it lacks. Her tools were primarily etymology and an attitude of valour towards the past as a territory in which poetically to dwell. On that mountain top, the artist and the intellectual stand tall, displaying an education that society no longer values; their classical etymology (*poiesis*)—although literally about making and creating—re-imagines for readers a different and more serene life, instilling it with the adventure of abstraction. The unchanging quality of the Greek word is part of its classical character—no longer open to revision since it is already perfect in itself. The artist-intellectual is 'poetic' insofar as he or she preserves the values of honest hopelessness and hungry erudition over the peasants of the office shift and their counterparts, the vulgar billionaires.

A second 'poetic' shares an attitude with the first, but lacks its philosophical cast of mind, equally at home in the past's serenity, but from a less insurrectionary angle. One thinks here of new modernism studies, which has produced in the last decade and a half a phalanx of books that re-assert the rights of the canonical modernist centre of English studies, which has held court since the Second World War. Offended by postcolonial studies, and caught off-guard by the unapologetically positivist turn to new sociologies of world literature—with its market republics, algorithmic data compilations, and world systems mappings—new modernism studies launches a rejoinder. It dusts literary modernism off, polishes its political credentials, and demonstrates its flexibility in providing all the insights into race, class, and empire that the postcolonial insurgents lent to the middle-class black diaspora. Here the poetic is the possession of the geniuses of small literary magazines and renegade bourgeois salons. Their icon is that of the rebel destroying form in order to master it, living unconventionally, travelling wildly, and creating a fictional maze for hiding. A literature of perpetual irresolution and dissatisfaction loves the seminar room, and makes sure the work of interpretation is never done.

But neither of these is our 'poetic'. It is not a quality of literary experience, first of all, not the thrill of rhyme, the ecstasy of lofty sentiments, the non- or anti-real, the heightened awareness of the contemplative moment; it is not what Ezra Pound called "the obscure reveries of the inward gaze" (lines 25–26). It is not referring to the transport of the literary dragged into the world of the everyday, annexed to it, and superimposed upon it. We are not talking about the utopian desire that Pierre Macherey once identified as literature's compensation for rotten existence, where social problems are worked out symbolically to balance the ledgers of not working them out actually.

These various meanings are not even the opposite of what Vico had in mind when he spoke of "poetic history," "poetic politics," and "poetic physics"; they simply do not touch it, remaining somewhere off to the side. In Vico's vast study of the past—a paradoxically historical account of human prehistory—abstraction is the villain. The time before history that he explores is not at all a time when nothing happens (the great migrations out of Africa, the mastery of fire, the invention of the wheel, the first animal husbandry, the giving birth to God, the first burying of the dead—these are not nothing!); it is a time when no written documents existed, and therefore a time that can only be recovered by the documents later drafted out of others' speaking, composing, and remembering. His point is *on the other side of* those who say that since prehistory is unknowable it can only be re-assembled through myth—that is, people telling stories about what they had no access to, unchaining their imaginations with wild tales that in the clear light of modern reason we recognize as mere "poetic histories"—cheering up the miserable with tales of a world of non-contradiction.

Vico, by contrast, argues that pre-history is accessible if we come to understand that 'myth' is logical and rigorously historical; that myths are not compensations for an irrecuperable past, but precisely the eye-witness record that early humans left behind. The 'poetic' is not a separable literary province of aesthetic ideology but an innate feature of the logic of the first peoples who, as yet incapable of abstraction, wrote in figures and images. The 'poetic' then is a feature of language per se—at least the language that was still alive with the sensuous world of things, of the tools of labour, of natural animals and plants, of crops, of the most immediate sensations of the body. Language before the time that complex forms of social organization—above all, the hieratic seclusion of intellectuals and artists from modes of basic material reproduction—made thinking more mediated and abstract, and created the desire for thought to purify language of its cruder tendencies in the quest for rational clarity. Vico's story, then, is about the actual historical facts of poetic language as 1) the first language of human beings struggling to distinguish themselves from beasts by instituting marriage, burial, religion, and law; 2) the feature of early language that, once understood, becomes a key for deciphering the historical content of 'myth', which can be seen as a chronicle of harsh physical labour, the inequalities and class conflicts that came along with the first accumulated wealth of agricultural communities; 3) the sign of the loss we experience in modernity as it moves from vividness and immediacy to cold-blooded abstraction, and how the stratification of intellectual life arrogantly assumes for itself a superiority that conceals its own ignorance about what the myths were always trying to tell us, as though its way of thinking was the only valid one; 4) a rendering of the story of the world's peoples in ensemble, about their joint contributions, about their lamentable sectarianism, about certain races, ethnicities, and religions falsely claiming that they were first and the best, and that it was they who taught everyone else. His story is about the commonality of the world's peoples, about the barbarism out of which humans freed themselves through struggles over land and labour, creating laws of protection for common rights, the relapse into barbarism in later eras of historical return—but then again, a return from the return: the potential, even the inevitability, of trying again, with the chance of getting it right next time.

The Vichian outlook, the whole package, is so obviously resonant with the current state of imperialist adventure, racial swagger, the religious recoil against globalization, and the drama of mass migration that it is simply astounding that more have not thought to learn from it—to follow the threads back, for instance, to Ibn Khaldun, or to move forward through them to Marx, Gramsci, Max Horkheimer, Simone de Beauvoir, Mahdi Amel, Hayden White, Eqbal Ahmad, and Ernst Cassirer (among others), if only to provide some relief from the Auerbachian recyclings, the Wallersteinian borrowings, and the replaying of Derrida in the field of world literature. There are reasons for students of the 'poetic' to take notice of what the contributors in this book are doing—apart, that is, from the mostly unrecognized *traditions* of Vichianism represented by the names above in a theoretical landscape dominated by his nemesis Spinoza (an exciting story not yet told that lies behind a great deal of twentieth-century Marxism and anti-imperial thought), and apart also from the present undeserved respect for the natural sciences as a model for the humanities (a position that Vico eloquently identified as a "paralogism" while setting forth reasons why the humanities are the only science worthy of the name). It is only in Vico that we get a theory of the politics of literature that is not a reading of politics *into* literature, or the making of every micro-struggle over language a pretentious political event. In Vico, the 'poetic' is right there at the birth of the state, embroiled in its civic life. The first laws were literally poems. Vico's attractions (I first learned about him from Edward Said, although it took me many years to understand him) are as likely to be misunderstood as the word 'poetic' is. Like many of the contributors to this volume, I was moved by Vico's theories of history, language, and ethnic strife to see in a new light the anti-imperialist movements of the twentieth century. This connection may seem at first a great leap. But, for specific historical reasons, anti-imperial movements in the last century found valuable resources in Marxism as well as in non-Marxist forms of socialism and anarchism, often explicitly citing Vico in this regard, and drawing lines between the movements. All were inspired to some degree by the Soviet example, which was a powerful case of victorious anti-imperialist independence. The self-identity of the Soviet Union, its *raison d'être* and *inter*-national message to the world, was from the start—as a matter of sheer self-preservation as well its desire to spread its influence—stalwart intransigence to imperialism. It is quite impossible, unless one is in thrall to the prejudices of liberalism (as so many students of the humanities in American and British universities are), to deny that resistance to imperialism was essentially Marxist, and that Marxism and anticolonialism (as well as anti-imperialism) were inseparable. It may seem pat to say this, since postcolonial theory was deliberately devised to deny this fact, but the historical record—including the contemporaneous statements by many of the key figures extolled in postcolonial studies itself—is very clear on the matter.

The point would not be to go back and recount the failure of that promise, or the many betrayals of hope, the insufficient follow-through, or the simple fatigue of resisting the empire for decades surrounded, embargoed, and invaded as the empire so resourcefully struck back. It is a complicated history from which one should not flinch—that is, either from the failures and betrayals or the heroism and insight. One might, I am saying, learn more by doing something else. And that would be to see Marxism—but not only that—as an expression of Vichianism in its own time, *mutatis mutandis*, just as Vico's ideas took a particular form in Michelet's histories, Sorel's theories of Violence, Cassirer's theory of symbolic form, and Joyce's ideas of historical return so central to *Finnegans Wake*. Not all of the points of contact hold, of course. But it would be a matter of seeing something vexed and controversial (because too close to us, and encrusted with cant and fear) in light of the historical long-view, and to appreciate the sheer longevity of its intellectual gestures and the productivity of its many encounters. Perhaps it would help here to give an actual example. The case of the strangely neglected (although of course not unknown) Henri Lefebvre, for example, is an interesting one. In *Critique of Everyday Life*, he scoffs at the view of progress put forth by the captains of industry: like some "theological or metaphysical Providence, the human species is slowly advancing like a well-drilled army along a pre-ordained path from barbarism to civilization" (words written in 1947) (228–229). Isn't this sentiment—the one Lefebvre is attributing to the bourgeois—what Marxism is usually accused of arguing in American academia? Then why is Lefebvre, a philosopher in the French Communist Party, attacking it? He goes on to point out that the Chinese peasant who must light his house with paraffin is not unlike the poor on the outskirts of Paris who must do the same—that 'modernity' is haphazard, poorly distributed, leaving the invisible ones behind whether in the metropolis or on the periphery. When he refers to the United States a few pages later, he writes "America . . . where imperialism is alive and well" (235). He explores, moreover, these political outlooks translated into a literary style. In a fascinating critique of irony (unlike his embrace of irony in his later book, *Introduction to Modernity*) he writes,

abstract culture not only supplies him [i.e. André Gide] with words and ideas, but also with an attitude which forces him to seek the meaning of his life and his consciousness outside of himself and his real relations with the world.

(238)

The 'cultivated' man forgets the social foundations of 'his' thought. When he looks for the secret of his behaviour and his situation in words and ideas that he has received from without, he imagines that he is looking 'deep into himself'.

(239)

This is a typical, not an anomalous, outlook of twentieth-century Marxism, both first- and third-world. And it is a sensibility, it seems to me, consonant with Vico, not only because of its reservations with the arrogance of a metropolitan progress that cannot imagine itself capable of a historical return, but its *un*-modernist desire for real progress—its view that progress is not a poetic illusion to be chucked over by cynical knowledge. This point is made very well in the essay on Aimé Césaire in the present volume, where we are reminded that Césaire's humanism is often skipped over by today's critics under the weight of the go-to ethos of modernist despair, where it is forgotten that Césaire was taking his lead from the likes of Vico, Michelet, Sorel, and Lafargue rather than Schopenhauer, Kierkegaard, or Blanchot. Lefebvre's awareness of imperialism was defining, his abreaction to metropolitan inequality absolute. But it is this stress on the active man and woman, the makers of judgement, the analysts of their own existence, who take responsibility for this unacceptable state of impudent metropolitan dominance, blind to the little pockets of third-world misery in the metropolis itself. That is what is most Vichian here—this attitude of intellectual worldliness where Lefebvre explicitly ties together the 'introversion' of the literary pose and the dilemma of fake cultivation.

The varied, consistently innovative, collection of essays in this volume shows me, at last, that this same spirit and sensibility is alive in the generation now coming onto the scene, and that it has carved out a space neither dwelling in the past, nor having recourse to the standard-issue gurus of theory. From the decisive and inexhaustible importance of dissecting the 'global Nietzsche' and his novel of performance, the volume has the range to be able to move to problems of the materiality of language and the spoken word in Irish theatre. What could be more Vichian than this problem of the oral traditions of a language that was, despite being unwritten, initially so materially *there*, so physical, that it consisted of passing frogs, bows, or ploughshares between interlocutors as a means of symbolic expression? The Cold War uses of Platonov could not be more on point with the scenario I just sketched above, of course, but the essay on Platonov goes a brilliant step beyond by showing how the linguistic turn of high theory was a way of *not* making one's political views explicit, remaining true to Vico's critique of the "conceit of the scholars." It is one thing to invoke out of sheer decency and inclusiveness the intellectual contributions of third-world intellectuals against the great odds London and Washington imposed on them, but quite another to document them in an unforced, non-segregating way. The welcome essay on Mao and Tagore as architects of an Asian vision of socialist universalism is, in this light, deeply consequential, not only for its story of a solidarity of ideas that needs no metropolitan intermediary, but because it demonstrates that humanism is indeed universal and that it—like Vico would have it—is an aspiration universally shared, in this case between the often hostile countries of India and China.

Remarkable in this collection is the consistent sensitivity to literary form. Although of ambitious social import, it is the tactile sensitivity of the essay dedicated to an unsung constant in peripheral aesthetics—here, the intriguing mockery of irony in the *Estridentismo* movements of Mexico on behalf of an art that is nonetheless knowing, informed, and, in a word, avant-garde. How lively it sits beside the quite different essay on José Enrique Rodó's concessive Uruguayan masterpiece *Ariel*. If so many writers from the Caribbean embraced the image of Caliban rather than Ariel—identifying with the illiterate servant rather than the sprite as a resistance to Shakespeare's original insult—Rodó's title turns the table on Caliban, as well as the image of 'borrowed light', by leaving the whole question of the present and the past, the first and the copy, to the side in the effort to avoid polemics. In a line of thought attentive to buried traditions, the important reconsideration of the Brahmo Samaj—the secularist Hindu movement of zamindari reformers—takes us skilfully out of the pinched frames of Subaltern Studies, and shows us an alternative tradition.

The dance between India, China, and Latin America in the collection continues with the salient attention paid to two apparent outliers: the almost forgotten Pakistan, for instance. In the essay dedicated to the unfathomably neglected work of Muhammad Iqbal, an invaluable piece of intellectual history explores the contested borderlands between British idealist philosophers and Iqbal, over their use of Hegel. Second, indigenous culture—neglected by many scholars, whatever their camp affiliations. In this essay, one is asked to consider by way of the extraordinary novels *Yawar Fiesta* and *Deep Rivers* the reality of the *indio* in the unique vision of José María Arguedas—its full-fledged, off-kilter experimentalism, and the unfairness of writing it off as a kind of ethnographic quirk of Latin American regionalism. But China speaks loudly in this volume yet again in a startling and informed rendering of the country as a former colony—against those who argue it never was—and the role within it of the great translator and short-story master, Lu Xun, who in philological fashion delivered a message of aesthetic worldliness in the unsystematic, prolific way that was Vico's manner.

An anodyne response to prevailing whiteness, postcolonial studies has tended to thrive in the little ghettos that are those 'also' courses in the curriculum—the electives reserved for the department's noble souls and ethical agents. It is one of the many consequences of the disastrous rejection of universalism within postcolonial studies—a universalism that might have forced José Vasconcelos into the company of Matthew Arnold, or made it natural to read Mao when reading Marcuse on contradiction. Meanwhile newspaper reviewers, in the otherwise welcome openings for writers from the former colonies, kept hunting for that deep epistemological otherness that one supposed could be found in the writer of another skin tone with a foreign-sounding name. In the classrooms, newspapers, and museum exhibitions, work from the periphery is for the most part dealt with out of context, given a fair-minded close reading perhaps, but left to migrate to the comfortable aesthetic home of the metropolis's modernist tastes, now trained on the edgy topics of race, difference, prejudice, memory, and injustice (topoi that enhanced the taste rather than the other way around). The 'I' of the author was first and last.

Vichian studies rather build from a human point, collective and largely anonymous, and they develop their complexities by way of ideas and texts rather than the docile body of difference. A book about an eighteenth-century Neapolitan, two nineteenth-century Germans, and a twentieth-century French eccentric—conducted, of course, with certain global ends in mind—has in this way been enriched and brought to completion by a volume like the present one entirely dedicated to non-European sources. It shows the poverty of the charge of Eurocentrism when it is based on the racial or ethnic identities of authors rather than the interests and outlooks of metropolitan exclusivism and dominance.

Position and location rather than inheritance are what matter most. The point would be not blithely discriminating, not forgetting the pockets of poverty and disenfranchisement in Lisbon, Wallonia, and Alabama along with Gabon and Myanmar; not pretending we are securely within our families rather than hopelessly embroiled in each other's business, as though a rigid East–West or North–South divide was even possible under imperialism. As though imperialism had not precisely obliterated that and forced us all to pick up the pieces to make something new.

Works cited

Lefebvre, Henri. *The Critique of Everyday Life*. Vol. 1. Translated by John Moore, Verso, 1991.

Macheray, Pierre. *A Theory of Literary Production*. Routledge and Kegan Paul, 1978.

Pound, Ezra. *Hugh Selwyn Mauberley*. 1920.

Vico, Giambattista. *The New Science of Giambattista Vico*. Translated from the third edition (1744) by Thomas Goddard Bergin and Max Harold Fisch, Cornell UP, 1948.

Acknowledgements

The idea for this anthology arose out of a memorable Historical Materialism conference panel at NYU in 2015. I would like to acknowledge the organizers and participants—and particularly Timothy Brennan—who invited me to serve as a discussant for the *Borrowed Light* panel. Neil Larsen provided valuable comments during the early phases of the book proposal, and Tim provided detailed feedback throughout the process. Thanks to Faye Leerink, Mark Jackson, and the three blind reviewers for their interest, critical comments, and gracious support. Thanks also to Ruth Anderson, Sally Evans-Darby, Colin Morgan, and the Routledge team. Finally, I am grateful to Ralph Pollock and William Hatherell for their help in preparing the manuscript. I would especially like to thank Ralph for his thoughtful and careful editorial assistance, and Alaa Dabboor for formatting support. This book is dedicated to Shazia, Armaan, Irsa, and others who embarked on this journey with me.

Index

Printed and bound by CPI Group (UK) Ltd, Croydon, CR0 4YY

24/10/2024

01778282-0011